ADELPHI

Paper • 297

Beyond Indochina

Contents

Oxford University Press, Walton Street, Oxford OX2 6DP
Oxford New York
Athens Auckland Bangkok Bombay
Calcutta Cape Town Dar es Salaam Delhi
Florence Hong Kong Istanbul Karachi
Kuala Lumpur Madras Madrid Melbourne
Mexico City Nairobi Paris Singapore
Taipei Tokyo Toronto
and associated companies in
Berlin Ibadan

Oxford is a trade mark of Oxford University Press

Published in the United States
by Oxford University Press Inc., New York

First published July 1995 by Oxford University Press for
The International Institute for Strategic Studies
23 Tavistock Street, London WC2E 7NQ

Director: Dr John Chipman
Deputy Director: Rose Gottemoeller
Assistant Editor: Juliet Sampson

British Library Cataloguing in Publication Data

Data available

Library of Congress Cataloging in Publication Data

ISBN 0-19-828024-6
ISSN 0567-932X

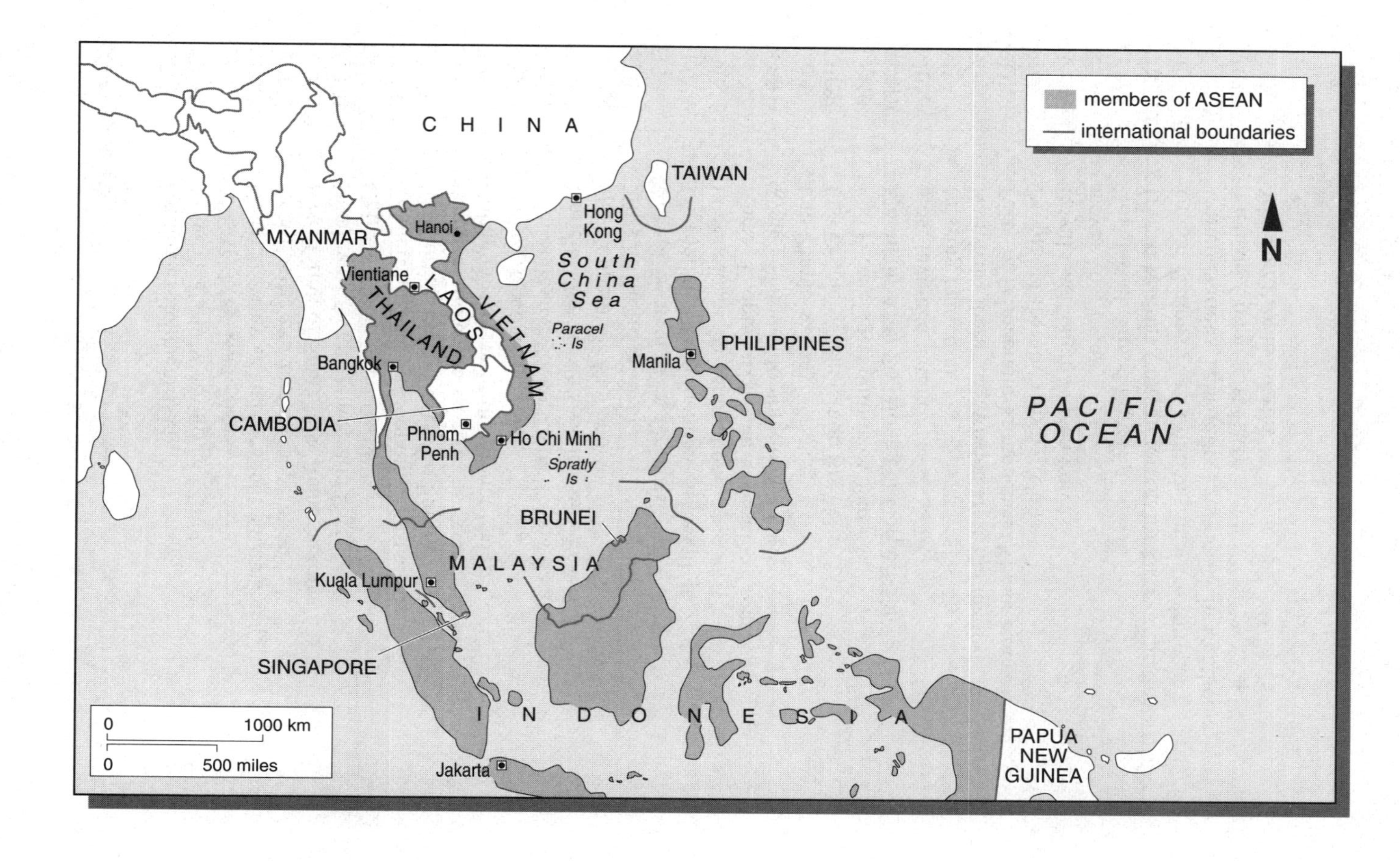
members of ASEAN
international boundaries
N
CHINA
TAIWAN
Hong Kong
MYANMAR
Hanoi
Vientiane
LAOS
THAILAND
VIETNAM
South China Sea
Paracel Is
PHILIPPINES
Manila
Bangkok
CAMBODIA
Phnom Penh
Ho Chi Minh
Spratly Is
PACIFIC OCEAN
BRUNEI
MALAYSIA
Kuala Lumpur
SINGAPORE
INDONESIA
PAPUA NEW GUINEA
Jakarta
0 1000 km
0 500 miles

GLOSSARY

ADB	Asian Development Bank
AEWAC	Airborne Early Warning and Control
AFTA	ASEAN Free Trade Area
ANKI	National Army of Independent Kampuchea
APEC	Asia-Pacific Economic Cooperation
ASEAN	Association of South-east Asian Nations
BLDP	Buddhist Liberal Democratic Party
CCP	Chinese Communist Party
COMECON	Council for Mutual Economic Assistance
CPAF	Cambodian People's Armed Forces
CPP	Cambodian People's Party
CPSU	Communist Party of the Soviet Union
EEZ	exclusive economic zone
ESCAP	Economic and Social Council Asia-Pacific
Funcinpec	United Front for an Independent, Peaceful and Cooperative Cambodia
GDP	gross domestic product
GSP	General System of Preferences
IDA	International Development Association
IMF	International Monetary Fund
KPNLF	Khmer People's National Liberation Front
KPRP	Khmer People's Revolutionary Party
LPA	Lao People's Army
LPDR	Lao People's Democratic Republic
LPRP	Lao People's Revolutionary Party
MADC	Mountain Area Development Corporation
MFN	most favoured nation
MIA	Missing in Action
Molinaka	National Liberation Movement of Cambodia
PDK	Party of Democratic Kampuchea (Khmer Rouge)
PLA	People's Liberation Army (China)
RCAF	Royal Cambodian Armed Forces
RTA	Royal Thai Army
RTAF	Royal Thai Armed Forces
UNCLOS	United Nations Convention on the Law of the Sea
UNHCR	United Nations High Commissioner for Refugees
UNTAC	United Nations Transitional Authority in Cambodia
VCP	Vietnam Communist Party
VPA	Vietnam People's Army

INTRODUCTION

Indochina has changed beyond all recognition. The collapse of the Soviet Union has removed the traditional source of military and economic aid to the Indochinese states. The war in Cambodia has ceased to divide the region into two antagonistic blocs. Military confrontation between Indochina and its non-communist neighbours has given way to cooperation and talk of economic integration. At the same time, Laos and Vietnam have jettisoned central planning and socialist economics, while Cambodia has discarded Marxism-Leninism. The opportunities for integration with the Association of South-east Asian Nations (ASEAN) and the creation of a new regional order have never seemed greater.

Two key factors are likely to shape the future of Asia's wider strategic environment: the shifting relationships between the major regional powers; and rapid economic growth and political change. The stability the Cold War imposed on the region may be replaced by a more fluid and complex environment, and a new strategic balance is likely to emerge in which the roles of China, Japan and India will change.[1] A new relationship may already exist between Indochina and ASEAN, even though no security balance within South-east Asia can be entirely independent of the larger East Asian region.[2] The challenge facing South-east Asia is to create an effective regional order based on incorporating the states of Indochina and Myanmar, and developing an effective security mechanism, such as the ASEAN Regional Forum, that will be able to institutionalise the new power balance. This Paper analyses how the states of Indochina are meeting the challenge and how its neighbours have reacted. It seems clear that as the states of Indochina increasingly go their own way towards integration with ASEAN they are likely further to complicate the already complex politics of South-east Asia.

But what is 'Indochina'? Historically the term has had several meanings. It has been a cultural region in mainland South-east Asia, a group of French colonies, a military theatre of operations during the First, Second and Third Indochina Wars, and finally, an alliance of socialist states led by Vietnam and backed by the Soviet Union.[3] Each of these definitions, in its own way, implied greater unity and coherence in this subregion than has proven to be the case. The collapse of the 'red brotherhood' in the 1970s, immediately after communist parties triumphed in Cambodia, Vietnam and Laos, illustrates this point.[4] The collapse led to the argument 'that Indochina is now only a category of convenience rather than a set of special relationships subject to Vietnamese dominance' and that 'the concept of Indochina as some kind of unitary actor needs to be radically revised'.[5]

Indochina as a politico-military bloc no longer exists. Trilateral coordination of foreign policy and economic and political relations has ceased.[6] The Cambodian peace process has taken that country completely out of Vietnam's orbit. Vietnamese forces have withdrawn from Laos, and Laos in turn has sought to develop a more balanced set of external relations. Nonetheless, after the Cold War, internal and external transition is by no means assured.

There are a number of fundamental uncertainties in this transition. Foremost among them is the effect of massive devastation of Cambodian society after years of internal war. The weakness of the Cambodian state and its difficulty in containing Khmer Rouge resistance illustrate the continued instability that could slow the transition process.

Significantly, Indochina is also host to the Vietnam Communist Party (VCP), one of the largest and oldest communist regimes still in power. All three Indochinese states have embarked on programmes of economic reform and political liberalisation and there is no guarantee that domestic transition will be trouble free. Economic and political change has already caused and inevitably will continue to bring about upheavals in society.

Thailand still harbours suspicions of Vietnam, as demonstrated by its reluctance, until very recently, to sever relations with the Khmer Rouge. Despite the normalisation of Sino-Vietnamese relations in November 1991, it is clear that underlying tensions remain. All of these factors could impede the forging of effective links between the states of Indochina and South-east Asia.

South-east Asia and the larger Asia-Pacific region, 'Indochina', may well become a historical concept. But can change proceed smoothly? What are the implications, for example, of Cambodia's long-term debilitation on regional integration and the creation of a new regional order? If the transition process is not successful, Indochina may well remain a zone of contention and therefore a significant geopolitical trouble spot.

A careful analysis of Indochina must treat the three states and their national interests as distinct: Cambodia, Laos and Vietnam pursue their own domestic agenda and foreign-policy priorities. The three states are not only at different stages of development, but are proceeding towards regional integration at different speeds.

Vietnam is a special case. With a population of 72 million, it is the second most populous country in South-east Asia and the twelfth largest in the world. Vietnam's population is roughly eight times larger than Cambodia's and 15 times that of Laos. Its natural and human resources, coupled with its drive towards a market-oriented economy have led some observers to view Vietnam as the next Asian tiger. It does have a potentially decisive role in the future of the region, and for that reason it is the major focus of this paper.

The 1994 construction of the first land bridge across the Mekong from Thailand into Laos makes possible the exploration and development of this landlocked country.[7] Additional land routes are being constructed between southern China, Myanmar, Thailand, Vietnam and Laos, as well as a road system between Singapore and southern China. As a consequence, Laos could move from the periphery to the 'heart of the Mekong'.[8]

The situation in Cambodia remains uncertain. The 1993 UN-held elections restored Cambodia's sovereignty and international standing but achieved little else. Cambodia's domestic instability poses security problems for each of its neighbours. In addition to the danger of internal conflict spilling over Cambodia's borders, domestic political actors might choose to exploit foreign policy for political ends. Alleged Thai support for the Khmer Rouge and the fate of ethnic Vietnamese are two issues already on the domestic political agenda.

Thus the three Indochinese states have very different prospects for domestic development and foreign relations. Their political stability, internal security and economic development will differ. Their future regional roles and alignments towards ASEAN and other regional bodies are also likely to be unique. Indochina is no longer a coherent geopolitical unit and each of its three states are separate actors with diverse national interests, pursuing different domestic and foreign-policy priorities.

I. THE DOMESTIC SETTING

Vietnam, Laos and Cambodia are changing from centrally planned to market-oriented economies. The political and economic changes in this process will in large measure shape regional order and relations.[1]

Mono-Organisational Socialism

The communist élite of Vietnam, Laos and Cambodia all embraced the precepts of Marxism-Leninism during their revolutionary struggles. Coming to power in North Vietnam in 1954, Vietnamese communists fashioned their own political order on the state constitutions and party statutes of China and the Soviet Union. The same model was applied to South Vietnam after reunification; Laos borrowed its system and political practices from Vietnam, and Cambodia imported them wholesale.

The supposed ideological similarities of the Indochinese states were reinforced during the Cold War when Vietnamese hegemony was at its height. The adoption of socialist vocabulary, forms and practices obscured fundamental differences among the states of Indochina on the one hand, and between the Indochinese states and their ideological allies in Eastern Europe and the Soviet Union on the other. The Leninist model barely penetrated the rural areas and the changes in the late 1970s were induced from the 'bottom up'.[2]

All three states experienced periods of non-communist rule during which domestic economies operated very much along free-market lines. Cambodia and ostensibly Laos were both monarchies in their post-independence years. South Vietnam (1955–75) and Cambodia (1970–75) were both republics in which the military was prominent. Now this heritage of non-communist political and economic organisation has reasserted itself. For example, Cambodia has reverted to its monarchical traditions while free-market forces have emerged most strongly in southern Vietnam and also Laos.

Political leaders in the Indochinese states attempted to adapt the Leninist model of 'mono-organisational socialism' to their societies. The essential features of this model were a one-party political system, and a command economy in which land and all major means of production were collectivised and subject to central state planning. The term 'mono-organisational socialism' described the party's attempt to monopolise political power.

In a Leninist system, all legal organisations and institutions are subject to party direction and control, including the highest organs of state; provincial, district and local government; the armed forces; and all other groups and associations. The party reserved the right to set the objectives of these groups, determine their structures and approve their leading personnel. It employed a system of dual or overlapping appointments to exercise control over state institutions (including the national assembly, council of ministers, ministries and central departments) and the military. For example, members of the party's Politburo and Central Committee also served as ministers in the government, generals in the armed forces or chairpersons of mass associations. Party members were organised into cells within these structures.

The outbreak of the Third Indochina War in the late 1970s resulted in the isolation of the Indochinese states from the Asia-Pacific region. This had two major consequences. First, Vietnam tightened its hegemonic grip on its two neighbours and coordinated their national security and foreign policies. Second, Vietnam, Cambodia and, to a lesser extent, Laos, became virtually dependent on the Soviet Union and Eastern Europe for trade, aid, investment and military support.

The war also coincided with the failure of the socialist economic system in Vietnam and Laos and this resulted in bottom-up pressures on reform. By the mid-1980s in Vietnam and Laos, and by the late 1980s in Cambodia it was clear that the ruling parties had been inept economic managers. Each regime had become dependent on external support – economic, military and political. One of the primary questions now facing the three Indochinese states is whether they can break out of this vulnerable position of dependency to stand on their own.

From Planning to Market

In the late 1970s, the socialist command economy failed in Vietnam and Laos. Both responded by adopting limited reforms, the most significant of which were in the agricultural sector. Laos rapidly disbanded its collectives, while Vietnam was more cautious. These initial measures proved too limited and pressures grew for a decisive shift to a market economy, although the armed conflict in Cambodia prevented the adoption of economic reforms at the time.

The second phase of reforms in Laos and Vietnam coincided with the rise of Mikhail Gorbachev and the implementation of his reform programme in the Soviet Union. During the mid-1980s, therefore, the reformist tendencies in Indochina and the Soviet Union were mutually supporting. In 1986, at landmark party congresses, both Laos and Vietnam codified their economic programmes under the slogans *pean pang mai* (new thinking) and *doi moi* (renovation), respectively.

The pace and scope of reform picked up considerably. Vietnam's sixth national party congress (December 1986) formally recognised the role of the private sector. After the congress, Vietnam created a 'multi-sectoral commodity economy' by dismantling the 'bureaucratic centralised system'. State subsidies were phased out or ended altogether in many areas. The managers of state enterprises were given greater autonomy. In 1988, a major price review freed the distribution and circulation of goods. The interest rate structure was made more flexible the following year, leading to an explosion of private entrepreneurial activity. Vietnam also promulgated a foreign investment law and invited capital investment from non-socialist economies.

In 1988, Vietnam decollectivised agriculture. Under Politburo Resolution 10 (April 1988) the peasant household was officially recognised as the basic unit of agricultural production. Although individual land ownership was not permitted, peasants were granted long-term rights of access to the land. As a result, agricultural production rapidly increased, making Vietnam the world's third largest exporter of rice, and rice became Vietnam's second largest export earner after oil.

Vietnam's *doi moi* programme embraced other important areas such as management, administration and commercial law. Initial reform efforts took several years to take root, partly because of the opposition of party conservatives. However, when the economic benefits became tangible, an increasing number of party members began to accept the reforms. Nonetheless, Vietnam still faces several major problems, not least is what to do with the large number of inefficient state-owned enterprises that are being kept afloat by indirect state subsidies. Vietnam has also moved very slowly down the road to privatisation and lacks the large sums of investment capital needed to improve its eroded infrastructure.

Economic reform in Laos paralleled that of Vietnam but was farther reaching. According to one specialist, Laos was the first socialist country successfully to attempt the 'big bang' approach to reform.[3] In July 1979, Laos suspended its programme to collectivise agriculture in the face of mounting evidence that it had been a failure. Party plenums in Vietnam (the sixth, August 1979) and Laos (the seventh, November 1979) formally approved the first cautious market-oriented reforms. In 1985, Laos introduced a 'one market, one price' system and stopped agricultural subsidisation.

National party congresses in Vietnam and Laos in late 1986 were major turning-points. The Lao People's Revolutionary Party (LPRP) congress approved the shift away from central planning towards the 'new economic mechanism'. In 1988, the role of agricultural cooperatives was reduced and peasants were granted long-term access to the land, decision-making in state enterprises was decentralised, and a foreign-investment law was adopted. Provincial check-points that hindered the flow of goods were abolished. In March 1990, Laos completely privatised state-owned enterprises. At the fifth congress in March 1991, Laos also officially recognised the role of the household economy and embarked on the wholesale divestment of state-owned enterprises, thus laying the foundation for market-based industrialisation.[4]

Cambodia's reform trailed behind its neighbours, largely because of its different socio-economic and security conditions. The initial reform programme dates to October 1985 when the fifth congress of the Khmer People's Revolutionary Party (KPRP) officially recognised the private sector (alongside the state, collective and family). In July 1988, the KPRP permitted private entrepreneurs to run businesses jointly with the state. Previously, all factories were government-run, although in practice the state turned a blind eye to small private workshops. Overseas Cambodians were encouraged to invest in joint ventures in Cambodia.

It was only in 1989 that Cambodia took determined steps to dismantle the remnants of central planning and move in the direction of a market economy. In April, the state Constitution was amended to recognise land-ownership rights and various types of land tenure. In July, Cambodia's first foreign-investment law was adopted, and after 1989, Cambodia began devolving control over state enterprises. A few were permitted to fix their own prices, and later, state enterprises were either leased to the private sector or sold for cash. The devolution of control was so rapid that it has been described as 'spontaneous'.[5] Cambodia's shift to a market economy was made easier by the small size of its industrial sector, the weakness of the state and the already existing large private economy. These reforms were accelerated in 1991 as a result of the impending Cambodian peace settlement and the need to amass domestic support in advance of national elections to determine Cambodia's future.

The result of economic reforms in all three Indochinese countries has been the abandonment of the socialist model of central planning. Indeed, the very 'definition of "socialism" has been modified beyond recognition', and an imperfect market economy has emerged in all three states.[6] State-owned enterprises are being commercialised and privatised (more so in Laos and Cambodia than in Vietnam). Agriculture has been decollectivised, and agricultural production, the largest contributor to gross domestic product (GDP), is now firmly in the hands of peasant families who have been given legal guarantees of direct access to, and long-term tenure of, farm land.

As economic difficulties mounted in the Soviet Union and Eastern Europe, two-way trade with Indochina began to collapse and aid dried up. Even traditional barter arrangements were phased out. Vietnam reoriented its trade away from the Soviet Union (with which it had conducted two-thirds of all trade) to Western industrial countries, Japan and the developing Asian economies. Both Laos and Vietnam also sought foreign investment from these non-traditional sources. Once again, the newly industrialising economies of East Asia became dominant.

The Cambodian peace settlement of October 1991 was a watershed in the development of Vietnam's foreign economic relations. Once the agreement was signed, regional states dropped their long-standing embargo on trade, aid and investment. In July 1993, the United States relaxed its policy of blocking loans to Vietnam by international lending agencies such as the International Monetary Fund (IMF), the World Bank and the Asian Development Bank. US law, however, prohibited foreign assistance to communist Laos and Vietnam. In February 1994, the United States lifted its 30-year embargo, although Vietnam is still denied access to trade and investment benefits accorded under most favoured nation (MFN) status, Generalised System of Preferences privileges and credits from the Export-Import (Exim) Bank of the United States.

These developments have several notable implications. First, in a relatively short period, Vietnam and, to a lesser extent, Laos reoriented their foreign economic relations away from the former Soviet bloc towards free-market economies. Ideology as a feature of foreign policy was greatly devalued. For example, Vietnam has established diplomatic relations with South Korea and Israel, and has developed close economic links with Taiwan. Laos has turned to Japan, Thailand and France.

Second, economic factors now have greater importance in the national security calculations of Vietnam and Laos. Since the seventh party congress in mid-1991, for example, Vietnam has held that a stable and peaceful external environment is necessary for Vietnam's economic development. Vietnam has moved, therefore, to 'multilateralise' its foreign relations, that is, to open its doors to all comers. Expanding commercial engagement with the world economy has led to new pressures on Vietnam to conform to international commercial and legal practices. Foreign trading partners and foreign investors want to minimise the risks of doing business in Vietnam, as the country's legal base is woefully inadequate. Vietnamese leaders have responded by modifying foreign-investment laws and regulations and by giving priority to commercial legislation. A similar process is under way in Laos.

Third, the political economy of Indochina is being rapidly reshaped as aid donors, foreign investors and international financial institutions make their presence felt. Southern Vietnam, particularly the area around Ho Chi Minh City, is becoming a powerful commercial counterbalance to the political centre in Hanoi. Other regional centres, such as Hai Phong and Da Nang, are emerging. New political patterns are forming at all levels as party bureaucrats, state officials and the new entrepreneurial classes form alliances and engage in joint ventures with overseas firms.

Cambodia is a prime example of 'opening up', leading to the penetration of party and state institutions by outside actors such as aid donors and international financial institutions. Laos has attempted to insulate itself by restricting the pace of change, which has been possible because the political élite is so small and development is at a low level. Inequitable distribution of wealth, corruption and strong foreign influence have all emerged as contentious political issues.

What is clear, however, is that economic reform programmes have had a profound impact on the pattern of the Indochinese states' external relations. Each of the three has abandoned the past practice of coordinating economic planning on a long-term basis. This has been replaced by new forms of cooperation with regional states, such as the Mekong Commission, a six-nation successor to Interim Mekong Commission of the UN's Economic and Social Council Asia-Pacific (ESCAP) headquartered in Bankok.

All three states have moved from almost total dependence on the Soviet Union and Eastern Europe towards establishing trade and commercial relations with a wide variety of states. Vietnam has forged economic ties not only with its immediate neighbours in South-east Asia, but with North-east Asia, Australia, Europe and North America. Laos has been careful to keep its relations in a smaller circle. Its economy is still dependent on development assistance, which is provided mainly by a combination of international institutions, the Asian Development Bank (ADB) and International Development Association (IDA), and foreign countries (Japan, Sweden, France, Australia and Germany). Cambodia, too, is heavily dependent on development assistance. Its central government faces a mammoth problem in attempting to regulate cross-border trade in timber and gems as well as foreign trade through Cambodia's ports.

From Marxism-Leninism to 'Market-Leninism'

Economic reform in Laos and Vietnam was accompanied by efforts at political liberalisation.[7] The upheavals in Eastern Europe in 1989 reinforced the commitment of the ruling parties in Laos and Vietnam to the continuation of one-party rule and gradual political reform. The system is shifting from Marxism-Leninism to 'Market-Leninism'.[8]

Prior to the political eruption in Eastern Europe, Vietnam and Laos had already begun to experience the political effects of their economic reforms. However, the collapse of communism in Eastern Europe shook the ruling parties in all three Indochinese states. Vietnam, for example, labelled Solidarity, the Polish trade union movement, a 'reactionary organisation' and denounced its August 1989 electoral victory as a 'counter-revolutionary *coup d'état*'.[9] *Pasason*, the LPRP's newspaper, declared 1989 'a nightmare year for socialism' in its 1990 New Year's Day editorial. The Party Secretary-General, Kaysone Phomvihane, said that developments were 'confusing' and called for 'vigilance' on the home front.[10]

The VCP, LPRP and KPRP reacted to events in Eastern Europe in late 1989 by reasserting that they would not give up their monopoly on power. They criticised calls for political pluralism, multi-party democracy and the 'depoliticisation of the military'. As if acting in concert, the VCP and LPRP both denounced the 'strategy of peaceful evolution' by which 'imperialist and other reactionary forces' were plotting to link up with dissident elements within Vietnam and Laos to overthrow the Communist Party by non-violent means. The strategy of peaceful evolution – a term used by party officials –has since become one of the prime national-security concerns of these states.

In 1990, Vietnam, Laos and Cambodia faced a new challenge mounted by individuals who advocated faster political reforms and who were critical of one-party rule. In all cases, the ruling party reacted by dismissing or imprisoning those concerned. In March, for example, Vietnam dramatically expelled Tran Xuan Bach from the Politburo for 'violating party discipline'. Bach had been one of the foremost advocates of political reform. In Cambodia, in September, a number of high-ranking state officials, including a minister, were jailed for attempting to form an opposition Social Democratic Liberal Party.

In Laos, also in 1990, a number of government officials, influenced by the pro-democracy movements in Eastern Europe, established an informal group of 'social democrats'. When the draft state Constitution was released for public discussion, they saw it as an opportunity to present their case for political reform. They argued that the article in the Constitution defining Laos as a 'people's democratic state under the leadership of the LPRP' should be dropped, and that Laos should become a multi-party democratic system. In August, they circulated their letters of resignation from the Party. In October, three senior government officials who called for multi-party democracy were arrested and detained until late 1992 when they were brought to trial.

All three Indochinese states held national party congresses in 1991. In Vietnam and Laos, the ruling parties reiterated their commitment to market-oriented reform and gradual political change. Despite calls by some party members for further political liberalisation, each party reaffirmed its opposition to pluralism and multiparty democracy and asserted its right to retain monopoly power.[11]

In both Vietnam and Laos, the aim of reform was to increase the efficiency and administrative capacity of the existing system. Emphasis was placed on more clearly demarcated roles of the party and the state. The party was to set broad policy goals while the state was to engage in day-to-day administration and policy application. The state bureaucracy was to be streamlined. There were attempts to make provincial government more responsive to central direction and therefore less able to block reforms.

Between 1989 and 1992, Vietnam wrestled with constitutional reform. No less than four drafts were considered before a revised state Constitution was agreed, which drastically amended that of 1980. The new Constitution recognises five economic sectors including the private sector; human rights and other freedoms have been written into the charter; collective leadership has been superseded by a system of ministerial responsibility; the Council of Ministers has been replaced by a cabinet government led by a prime minister, while the Council of State has been replaced by a state president.

In Vietnam, a major effort was made to make the National Assembly more effective as a legislative body. The electoral law was amended in 1992 to require every National Assembly seat to be contested, and for the first time it permitted independent candidates to nominate themselves. The Vietnam Fatherland Front and its affiliates were given greater responsibility for vetting and selecting candidates.

The July 1992 national elections disappointed those who sought to reform the system.[12] Many of the incumbent deputies who had been critical of government policy were dissuaded from running. Most independent candidates were disqualified during the pre-selection process, and the two who eventually stood were defeated. It was only in the urban areas, especially in Saigon, that voters had a relatively free choice. Some high-level party leaders failed to top the voting results, while some party-endorsed candidates were not elected. The National Assembly that resulted from the elections comprised a younger and much more educated group of deputies than its predecessors.

The political reform process in Laos paralleled that of Vietnam. In 1988, Laos held elections for district and provincial people's committees – the first elections since the establishment of the Lao People's Democratic Republic (LPDR) in December 1975. They were followed in March 1989 by national elections for an expanded Supreme People's Assembly. The composition of the deputies elected revealed continued party dominance and an over-representation of lowland Lao. The new Assembly set up a constitution drafting committee and created two new ministries: Economic, Planning and Finance, and Foreign Economic Relations and Trade. The government bureaucracy was also reduced by one-third.

In August 1991, Laos adopted its first written Constitution. The role of the LPRP as the 'leading organ' was spelled out in Article 3, but was not otherwise mentioned. Likewise, there was no reference to socialism, even as a distant objective. The hammer and sickle and five-pointed red star were dropped from the state's emblem. Top government structures were reorganised and renamed. Other important changes included enhancing the power and role of the renamed National Assembly, ascribing ownership of the land to the 'national community' rather than the state, and creating an independent judiciary. Elections to the National Assembly under the new Constitution were held in December 1992.

The situation in Cambodia was radically different. The KPRP held its congress on the eve of the signing of a comprehensive peace settlement, which included the institution of a liberal democratic political system through UN-supervised elections. At the extraordinary KPRP congress of October 1991, the Party declared a formal end to communism (by eliminating Marxism-Leninism) in favour of multi-party politics and a free-market economy. The KPRP also renamed itself the Cambodian People's Party (CPP). During 1992–93, the UN Transitional Authority in Cambodia (UNTAC) exercised supervision and 'control' over Cambodia's domestic affairs, including the electoral process.

In May 1993, UN-sponsored elections, contested by 20 political parties, failed to produce a clear winner. Of the 120 seats in the new Constituent Assembly, 58 were held by the National United Front for an Independent, Peaceful and Cooperative Cambodia (Funcinpec), 51 by the CPP and the remainder by two smaller parties. No party had an absolute majority let alone the two-thirds majority needed to adopt the Constitution.[13] The result was a fragile coalition government that agreed to create a constitutional monarchy headed by Norodom Sihanouk. UN forces effectively withdrew in late 1993.

Cambodia's political system is far from a liberal multi-party democracy. Both Funcinpec and the CPP are beset by internal factionalism, and there is antagonism between party representatives in the coalition government. Funcinpec's party structure, especially in provincial areas, appears to be disintegrating. The dominance of the CPP over the central government – including key ministries, the armed forces, the police, security services and provincial government – remains unchanged. While the CPP may have shed its Marxist-Leninist ideological trappings, it retains the old organisational structure.

The liberal political atmosphere created by the UN has been eroded. The coalition government has survived an attempted coup in July 1994. A reformist Minister of Finance who made strenuous efforts to end corruption in the customs service and increase central government revenues was dismissed from office in a rare display of unity by coalition partners. He was later dismissed from the National Assembly.

The fragility of the coalition government arises not only from factionalism and inter-party rivalry, but also from the state's fundamental weakness after years of warfare and the continued military challenge of Khmer Rouge guerrillas. Cambodia's democratic gains could easily be lost in a return to one-party rule.

Clearly the Leninist model is no longer a useful analytical device for understanding the political systems in Indochina. The mono-organisational socialism model may have explained the élitist politics in urban areas, but it can no longer account for the nature and scope of political activity that resulted from the economic reform process. The Leninist model exaggerates the political similarities between Laos and Vietnam when, in fact, they are different in important ways.

As a result of the opening up of Indochina in the late 1980s and the development of outward-looking market economies, there has been an explosion of new social groups. This is particularly noticeable in Vietnam and Cambodia, but less so in Laos. The growth of these social forces and the creation of a new political economy have profound implications for the future stability of each Indochinese state. While it is clear that they are no longer Marxist-Leninist, it is not clear what each is becoming. As the case of Cambodia demonstrates, traditional political forms that pre-dated mono-organisational socialism may reassert themselves. But there are other possibilities. David Marr has argued that in Vietnam there are three alternatives:

> If the Party decides to reassert proletarian dictatorship on these organisations, many are likely to continue operating from the shadows ... If the Party allows groups to proliferate and link up with each other, thus infiltrating the state more effectively, the political system will become more vulnerable to schism and protracted instability. The third alternative is to legalise the public sphere independent of either state or private sectors. Only then might one be able to talk about a 'nascent civil society'.[14]

Börje Ljunggren has suggested a fourth alternative – that Vietnam may evolve into a 'soft authoritarian system' following the pattern of South Korea and Taiwan.[15]

II. NATIONAL SECURITY

During the Third Indochina War (1979–89), Vietnam, Laos and Cambodia formed a unified military bloc to meet a common external threat. All the Indochinese states were dependent on the Soviet Union and Eastern Europe for military support. In addition, Vietnam was allied to the Soviet Union through a 25-year Treaty of Friendship and Cooperation. Now the strategic basis for their national-security policies has changed fundamentally. The three states no longer share a common threat, and their security concerns are primarily internal.

The subregional security relationships forged during this war, like political and economic relations, have also changed beyond recognition. Indochina was once a tight military bloc, but now Laos and Vietnam are exploring wider security arrangements, like the ASEAN Regional Forum. Cambodia is required by its Constitution to pursue a 'neutral and non-aligned' foreign policy. After several years of equivocation, Cambodia has decided to take the first step towards ASEAN membership by acceding to the ASEAN Treaty of Amity and Cooperation. This chapter examines how the three states are formulating new directions for their national-security policies.

The Death of the Indochinese Bloc

Vietnamese strategists have argued that Indochina formed a 'single strategic battlefield' and that the defence of Vietnam was inextricably linked with that of Laos and Cambodia.[1] In 1975, after communist-led military forces swept to power in South Vietnam, Cambodia and Laos, Vietnam attempted to formalise security relations with its neighbours. Laos signed a 25-year Treaty of Friendship and Cooperation in 1977, while Democratic Kampuchea resisted Vietnam's overtures. Formal security relations with Cambodia only became possible in 1979 when Vietnam invaded and installed sympathetic Cambodian officials.[2]

Ideology has also played a major role in shaping traditional national security doctrines among the Indochinese states. Throughout the 1950s, 1960s and 1970s, the communist parties in Indochina saw global politics as essentially falling into two camps. They viewed themselves as progressive socialists, led by the Soviet Union, which opposed the reactionary imperialists, led by the United States. In the late 1960s and early 1970s, Vietnam's analysis of world politics was based on the ideological framework of 'the three revolutionary currents', a Soviet ideological formation describing the correlation of forces and balance of power determined by the strength of the

Soviet-led socialist camp; the international communist and workers' movement; and the national liberation movement.[3]

From the Third Indochina War emerged a unified bloc of Indochinese states backed by and dependent on the Soviet Union and Eastern Europe. Military policies were among those coordinated across the three states. Despite the appearance of unity, the 'bloc' was never a firm security edifice. There were always underlying tensions, the most potent of which was Khmer nationalism. Cambodia's party chief, Pen Sovan, was reportedly removed from office because of his dislike of Vietnamese dominance. Laos never fully shared Vietnam's enmity towards China. In sum, there were fissures in Indochinese security policy from the start.

Gorbachev's Shadow

In the mid-1980s, the national-security assumptions of the Indochinese states were shaken by external developments.[4] Foremost among the developments was the 'new political thinking' in Soviet foreign policy. President Mikhail Gorbachev's initiatives in settling regional conflict by political means and normalising relations with China set in motion a major change in the South-east Asian strategic environment. This resulted in the 'de-ideologisation' of Soviet foreign policy, which turned Indochinese foreign policy upside down.[5] Of course, ideology still had a role in the normalisation of Vietnam's relations with China and, to a certain extent, it continues to shape the world view of Vietnam's leaders.[6] But in essence, Vietnamese national-security policy is now based much more on *realpolitik* than it had been.

Gorbachev's attempts to restructure the Soviet economy also had implications for Indochina. All three states carried out extensive market-oriented reform, including encouraging foreign investment through 'open-door' foreign policies. As a result, Vietnam, Cambodia and Laos each adopted a less ideological and more multi-dimensional approach to national security.

In late 1986, both the Lao and Vietnamese ruling parties embarked on far-reaching reform programmes. National-security policy was absorbed into this larger reform effort. The basic objective was to end socialist central planning by shifting to a market economy, thus integrating Vietnam, and to a lesser extent Laos, into the global economy. In order to make this shift, which depended on massive foreign investment, Vietnam had to overcome international hostility, particularly from China, arising from Vietnam's invasion and occupation of Cambodia.

Internal Threat

Vietnam and Laos both identify themselves as among the last bastions of socialism in the world. They allege that, along with China, they are the targets of the so-called 'plot of peaceful evolution' – a phrase that occurs repeatedly in their domestic media. The policies of these states declare that 'imperialists and international reactionaries' seek to eliminate one-party rule by encouraging domestic dissident groups to advocate political pluralism and multi-party democracy. Thus any suppression of domestic dissidents is justified on national-security grounds.

Vietnamese Politburo member and Minister of National Defence Doan Khue has described the strategy of peaceful evolution as the

> combined use of unarmed and armed measures against us to undermine in a total manner our politics, ideology, psychology, way of living and so on, and encircling, isolating and destroying us in the economic field, with the hope that they could achieve the so-called 'peaceful evolution' and make the revolution in our country deviate from its course. They have been trying to seek, build and develop reactionary forces of all kinds within our country; at the same time to nurture and bring back groups of armed reactionaries ... and to combine armed activities with political activities, hoping to transform the socio-economic crisis in our country into a political crisis and to incite rioting and uprisings when opportunities arise. They may also look for excuses to intervene, to carry out partial armed aggression, or to wage aggressive wars on various scales. Our people thus have the task of dealing with and being ready to deal with any circumstances caused by enemy forces: peaceful evolution, riot and uprising, encirclement, blockade, surprise attacks by [enemy] armed forces, aggressive wars on various scales. The politico-ideological front is a hot one.[7]

Laos makes a similar threat assessment, but internal political dissidence has dropped off. Repression of party dissidents and returned students in the late 1980s and early 1990s appears to have had its desired effect. Such is not the case in Vietnam. Every year the authorities publicise several court cases involving individuals charged with attempting to subvert the established order.

The main national-security problems for Laos in the 1990s are border security and internal insurgency. In recent years, Laos has placed priority on reaching frontier control and border-delineation agreements with its neighbours. Laos is primarily concerned about unauthorised border crossings along the Mekong, smuggling from Thailand, and unregulated settlement of Vietnamese in the southern provinces. A lesser objective is to prevent disputes over undemarcated territory from flaring up into armed conflict, as happened in the late 1980s. Finally, Lao officials are apprehensive that the opening of the Friendship Bridge across the Mekong will have a negative impact on Lao society. Specific concerns are smuggling, especially of narcotics, prostitution and the spread of AIDS.

Officially, the Khmer Rouge has been identified as the main threat to Cambodian national security. The draft reform plan of the Royal Cambodial Armed Forces (RCAF) General Staff, for example, sets as its priority the defeat of the Khmer Rouge within two years. Additional national-security concerns are increasing lawlessness and violent crime, border delineation and the status of Vietnamese residents.

Vietnam no longer faces a serious security threat from armed domestic insurgents, such as ethnic minorities or remnants of the former regime backed by Vietnamese living abroad. Political stability is not challenged by an organised opposition, although the potential for a clash with militant Buddhists belonging to the Unified Buddhist Church is a cause for concern. Nonetheless, the possibility exists that political turmoil could result from socio-economic change.

Shaping National-Security Policy in Vietnam and Laos

Indochina's strategic environment is shaped by a number of factors, primarily its geography and the military disposition of adjacent states. Indochina only shares land borders with China, Thailand and Myanmar, and the three states share borders with each other. Maritime boundaries extend the circle to include Indonesia, Malaysia and the Philippines.

In Vietnam and Laos national-security policy is virtually analogous to regime security: its central concerns are maintaining a single party in power and guaranteeing the independence and sovereignty of the nation-state. National-security threats – both internal and external – are viewed within this framework. The national-security perspectives of Vietnam and Laos contrast with the situation in Cambodia, where the unstable coalition regime is in danger of collapsing due to internecine rivalries. The Cambodian state is also weak, poorly developed and unable to exert administrative control over large areas.[8]

In the mid-1980s, an accumulation of internal and external factors radically changed national-security policy thinking. These were: a crisis in the domestic economy; an international trade and aid embargo; the costs of conflict in Cambodia; Chinese politico-military pressures; and Soviet insistence that the Indochinese states readjust to new political thinking. For example, as the Gorbachev reforms took hold in the Soviet Union, Laos and Vietnam began to refashion their national-security frameworks with greater emphasis on economic and technological factors.

In mid-1987, Vietnam adopted a significant new national-security doctrine embodied in Politburo Resolution 2, 'On Strengthening National Defence in the New Revolutionary Stage', which completely overturned the security orientation of the previous decade.[9] The past focus on the forward deployment of forces in Laos and Cambodia was replaced by an inward-looking, defensive policy. Vietnam withdrew its combat units from Cambodia and Laos. Main-force troops along the Sino-Vietnamese border were ordered to take a non-provocative stance, and as tensions declined their numbers were reduced. At the same time, Vietnam began a massive demobilisation of its main forces. An estimated 600,000 regular soldiers, including 200,000 officers and civilian defence specialists, were released from service by the end of 1990.

The tenets of Vietnam's national-security policy were endorsed at Vietnam's seventh national party congress, which met in June 1991.[10] They were incorporated into the 'Guidelines and Tasks for the 1991–95 Five-Year Period' issued by the VCP Central Committee's Military Commission. Since then, Vietnam has attempted to integrate all aspects of Vietnamese society – military, internal security, economic, social, cultural, political and foreign relations – into a single policy, embracing the 'people's war and all people's national defence'.[11]

Vietnam's defence doctrine was designed to meet what was perceived to be the most likely security threats in the 1990s; high on the list were low-level or limited conflicts. These were liable to occur in disputed off-shore territories in the Eastern Sea (South China Sea) or along Vietnam's land and sea borders, with China viewed as the main protagonist. Vietnam is also concerned about possible border incidents with Cambodia and clashes at sea involving Thai fishing vessels. In order to prepare for such conflict, strategic areas have been designated special defence zones. In the initial stages of limited conflict, the responsibility of meeting and countering an external attack will rest with the local militia and self-defence forces. At a later stage, the main forces, supplemented by ready reserves, would be employed.

Vietnam's much-reduced military establishment now places responsibility for meeting external threats in the hands of a mobile regular standing army equipped with the best weapons that Vietnam possesses, and backed up by a large strategic reserve composed of newly demobilised soldiers. In local areas, defence is the responsibility of militia and self-defence forces. These forces are also assigned to the Ministry of Interior to maintain public order and internal security. Each military region, province, city, district, ward and village is responsible for fortifying strategic defence zones such as land and sea borders, including offshore islands.

Laos has not made public an explicit national-security doctrine. This is partly because as a small landlocked state, it does not wish to antagonise any of its five immediate neighbours: China, Myanmar, Thailand, Cambodia and Vietnam. The Laotian armed forces are small and poorly equipped compared to those of its neighbours.[12] Laos' strategic environment is necessarily land-linked; historically, it has been invaded by the armies of neighbouring states, Thailand and Vietnam in particular.

Throughout the course of the Lao Revolution (1945–75), and more particularly, since the establishment of the LPDR in December 1975, Laotian national security has been based on a firm alliance with Vietnam.[13] Vietnamese military forces were stationed in Laos continuously from the late 1940s to the mid-1950s, and then from the early 1960s to the late 1980s.

During the Second Indochina War in the 1960s, Vietnamese forces engaged in offensives during both wet and dry seasons with the Pathet Lao, and maintained the security of the Ho Chi Minh Trail. After the LPDR was established, they also assisted in maintaining internal security. In July 1977, Laos and Vietnam signed a 25-year Treaty of Friendship and Cooperation, which called for consultations in the event of an external threat and legalised the status of Vietnamese forces in the country. Vietnamese forces deterred both Thailand and China during the Third Indochina War when Laos sided with Vietnam.

In recent years, as regional tensions have relaxed, Laos has perceived its security environment to be benign. This was first reflected in the LPRP's Political Report to the fifth party congress in March 1991. There was less discussion of defence and security compared to previous reports. National defence focused on maintaining 'peace and order ... national unity and solidarity and concord among tribes in the national community'. Only passing mention was made of possible external threats.[14]

According to Secretary-General Kaysone Phomvihan:

> In building armed forces for national defence and maintenance of peace and order, we must stress quality. Special significance must be attached to the building of regional armed forces who are capable of defending their respective localities.

In line with similar developments in Vietnam, the Lao People's Army (LPA) was directed to involve itself in economic production. As Kaysone noted:

> The national defence and security forces ... must promote the factors for self-strengthening, produce their own logistics supplies, partly fulfil the internal demands of their units, and contribute to socio-economic development.[15]

Shaping National-Security Policy in Cambodia

The rapidity of changing national-security perspectives and alignments in Indochina is best illustrated by Cambodia. Between 1989 and 1993, the government in Phnom Penh underwent three major changes before becoming a constitutional monarchy.[16] The one element of continuity has been the Khmer Rouge, still Cambodia's main security threat.

As a result of the adoption of a state Constitution and the formation of a coalition government in 1993, a national army – the Royal Cambodian Armed Forces – was formed by merging the National Army of Independent Kampuchea (ANKI) and Khmer People's Revolutionary Party (KPRP) military units with those of the Cambodian People's Armed Forces (CPAF). The actual size of the new army is disputed. In the post-election period, the UN provided the salaries for 128,000 personnel, which later crept up to 160,000.[17] In July 1993, Cambodian First Prime Minister Norodom Ranariddh put the strength of the armed forces at 130,000, while according to a 1994 assessment, 'RCAF strength on paper is about 130,000 but a more realistic figure would be less than 100,000, the difference being accounted for by "ghost soldiers" who remain on the payroll to help boost the income of unit commanders'.[18]

The new commanders are supposed to oversee the merger, reorganisation and demobilisation of surplus military units, and deal with urgent matters such as the payment of back salaries; adoption of a common uniform, insignia and badges of rank; provision of housing and social welfare for soldiers, demobilised troops, war invalids and their families; and standardisation of procedures and training. The new national army has taken charge of the de-mining programme, and the storage, maintenance and/or destruction of surplus equipment, weapons and ammunition.[19]

While reorganisation and restructuring were under way, the new Cambodian armed forces were tasked with maintaining internal security. From the outset, the newly established government and its military were beset by violence. The Khmer Rouge has consistently refused to observe a cease-fire, disarm and open areas under its control to the central government; it continues to instigate violence in a number of provinces, most notably Siem Reap, Banteay Meanchey, Preah Vihear and Kampot.

The Phnom Penh government was forced to declare that it had 'no other choice than to take firm measures to actively exercise the right to self-defence'.[20] Norodom Ranariddh called on the military to eliminate Khmer Rouge forces. In August 1993, RCAF units comprising forces of the newly merged factional armies launched offensive operations in Preah Vihear, Siem Reap, Banteay Meanchey, Battambang, Kompong Chhnang, Kompong Cham and Kompong Thom provinces. These actions sparked a record number of defections from Khmer Rouge ranks, including senior officers.

In 1993–94, the RCAF unsuccessfully attempted to seize and hold two major Khmer Rouge strongholds. RCAF forces mounted an initial thrust against Anlong Veng in October 1993 and captured it in early February 1994. Almost immediately an RCAF strike force of 7,000 personnel was directed to capture Pailin, an important Khmer Rouge gem-mining centre. In response, the Khmer Rouge mounted widespread diversionary attacks before counter-attacking and driving out government forces. Anlong Veng was retaken by the Khmer Rouge on 24 February 1994, and at least 10,000 civilians were displaced from their homes because of the fighting. In May 1994, foreign observers estimated that the Khmer Rouge was in control of more territory in Battambang and Banteay Meanchey provinces than before the elections, held a year earlier.

The upsurge in fighting led the Khmer Rouge to renew its demands for the formation of a quadripartite army and for its representation in the government in an advisory capacity. Prompted by King Sihanouk, the government explored possibilities for negotiation. Talks to discuss 'national reconciliation' were held in Pyongyang and Phnom Penh in May and June 1994, but these proved inconclusive. In June, after the Ministry of Interior stated that it could no longer guarantee the safety of Khmer Rouge representatives in the capital, they departed for their jungle headquarters. The government then officially outlawed the Khmer Rouge, offering amnesty to defectors. The Khmer Rouge responded by declaring the formation of a provisional government and increasing harassment.

The Cambodian government suffered a great loss of prestige as a result of these military set-backs, and morale in the armed services plummeted. The ineptitude of the CPAF, with its top-heavy structure, was exposed. The government's position was made even worse later in the year when local Khmer Rouge forces began capturing Western hostages and demanding ransoms for them.[21] Cambodian officials called on other countries, such as France, the United States and Australia, to provide military assistance. The Khmer Rouge promptly demanded the cessation of foreign military aid as a precondition for the release of the hostages. After the hostages' deaths were confirmed, the Khmer Rouge issued threats to the nationals of any country providing military assistance to Cambodia.

The failed government offensives of 1993–94 and the hostage incidents have highlighted poor troop morale and weaknesses in the RCAF's command structure and logistics organisation. The reputation of these forces has been further tarnished by a UN report revealing that a military intelligence group in Battambang was involved in torture and other human rights violations. The General Staff of the RCAF has unveiled a major reform programme that would pare down the military and ensure more professional standards.[22]

The formulation and implementation of Cambodia's national-security policy is obviously dependent on domestic circumstances. An important factor is the health of King Sihanouk, whose death is likely to cause political instability. The present government is a fragile coalition. Rivalries and suspicions between the CPP and Funcinpec have not entirely abated, particularly at the provincial government level where the CPP political structure is still in place.[23] While there are signs that Norodom Ranariddh and Co-Prime Minister Hun Sen have been able to forge a working relationship, its durability cannot be taken for granted. Each leader presides over a party structure splintered by cliques and factions.[24] Cambodia could revert to a single-party dictatorship, military rule or prolonged internal war, any of which could invite foreign intervention. In short, the continued debilitation of Cambodia acts as a brake on regional integration with ASEAN, and especially limits the possibility of Cambodia's integration into regional affairs.

The Military and the Economy

The collapse of the Soviet Union meant that the Indochinese states were suddenly cut off from an assured supply of military equipment. None had a modern defence industry to manufacture weapons, parts and ammunition, and none had the hard currency to purchase them. In Vietnam and Laos, priority was placed on storing and maintaining existing inventories.

In Cambodia, CPAF equipment deteriorated in combat and because of importation restrictions on military supplies mandated by the peace agreements. Although the RCAF, like its predecessor the CPAF, earns extra income from exporting timber, without significant external financing it will be difficult to maintain even the current low level of effectiveness of the Cambodian armed forces.[25]

Laos' defence posture was also partly determined by material backing from Moscow, which enabled it to establish a modern air force. Vietnam, whose military aid programme was always modest, could not fill the vacuum left by the Soviet Union. Laos approached China, but was offered arms on a commercial basis only. As a result, the armed forces were reduced by 18,000 during 1987–92, and the Lao Air Force has had to curtail drastically its air operations due to shortages of spare parts and aviation fuel.

Without external assistance, the capability of the Lao armed forces rapidly deteriorated as equipment was damaged and wore out. In order to meet operating costs, the Lao Army has also had to turn to commercial activities.

> The military ... therefore has turned to narcotics trafficking [to raise funds]. The military also runs several enterprises, the largest of which is the Mountain Area Development Corporation [MADC] which operates in Laxsai and Boulikampsai and sells logs to Japan, Taiwan and Korea. Military organisations are permitted to sign contracts with foreign investors. The MADC is also involved in drug protection.[26]

Vietnam's new defence doctrine also necessitated major changes on the home front, including a greater role for the military in economic activities. Regular forces were expected to become partly self-sufficient by growing crops and raising animals. Other military units were specifically assigned to production tasks.

In March 1989, Directive 46 of the Council of Ministers decreed that these army enterprises were on the same financial and legal footing as state-owned enterprises. Nine major Vietnam People's Army (VPA) companies were converted immediately into either corporations or general corporations. The most notable military enterprise was the Truong Son General Construction Corporation. While its main activity was infrustructure construction, it soon branched out into exporting coal, marble and coffee beans, laying railway tracks and general support services. In 1991, the Corporation was assigned a major role on the North–South powerline, the largest construction project since the Ho Chi Minh Trail. It now employs 7,000 people in 19 enterprises.[27]

Vietnam's national defence industries now manufacture a wide range of items in addition to military weapons and ammunition, including electrical and mechanical goods, bicycle parts, oil-burning stoves, electricity meters and fluorescent lightbulbs. Other army enterprises assemble television sets and radio-cassette recorders, manufacture garments and process agricultural goods. By 1993, approximately 12% of the standing army was employed full-time in over 300 commercial enterprises. A year later, the military had formed 26 corporations that were authorised to enter into joint ventures with foreign partners.

Despite the massive reduction of main-force regulars, Vietnam still maintains the largest ground army in South-east Asia, although it remains a 'poor man's army'. It is best suited to the regional defence of Vietnam and to low-level contingencies around its borders. The VPA has the capability to intervene successfully in Laos and Cambodia, but not elsewhere in the region. A weak industrial base and the lack of sufficient petroleum, oil and lubricants supplies would hinder its ability to conduct or sustain large-scale operations. Nor does it have the manufacturing capacity to produce modern armaments.

The Vietnamese Navy is primarily a coastal-defence force that has been assigned the added roles of exclusive economic zone (EEZ) surveillance (in the South China Sea and Gulfs of Thailand and Tonkin) and protection of the Spratly Islands. It has limited power in the South China Sea and farther afield. Vietnamese forces would be overwhelmed in any conflict with Chinese naval forces or combined ASEAN navies. Vietnam would also be at a disadvantage at the 'engagement level' in a conflict with Indonesia, Malaysia or Singapore; it could probably inflict serious losses only on the Philippines.

As early as 1988, Vietnam's Air Force was suffering from 'ageing aircraft and a critical shortage of spare parts', and the situation has since worsened.[28] The combat efficiency of the Vietnamese Air Force has undoubtedly been affected by the retirement of pilots with combat experience in the Vietnam War (1965–73) and by training restrictions imposed by fuel costs and shortage of munitions. Pilot proficiency has dropped and staff morale is low. It is reasonable to conclude that Vietnam has lost whatever qualitative edge it had over its rivals' personnel two decades ago.

Vietnam's Air Force is equipped for a self-defence role; it cannot mount strikes beyond Indochina, Thailand and southern China. It lacks in-flight refuelling and airborne early-warning aircraft (AEWAC) capabilities, and in its present condition, it could probably not maintain continuous air cover over the Spratly Islands. In any conflict involving the use of air power, Vietnam would be hard pressed to make up for battlefield losses in equipment and trained manpower due to its lack of an industrial base and shortage of trained pilots. The reported acquisition of six *Sukhoi* Su-27 fighter-bombers in 1995 will not redress this problem.

Over the last half decade the Indochinese states' equipment has deteriorated through ordinary wear and tear, lack of proper maintenance, and limited funds for spares and other major acquisitions. A technological gap is rapidly widening between Indochina and its more affluent South-east Asian neighbours, which have been engaged in an arms build-ups and the modernisation of their armed forces since the late 1980s. A net assessment of the military forces of the three states must be that they represent no credible conventional threat to their immediate neighbours.

The Indochinese states no longer equate national security predominantly with military strength. Their security policies are now more multi-dimensional, and are overwhelmingly domestic in focus. While all three have external security concerns, they no longer share the perception that they are threatened militarily by a common enemy. ASEAN no longer views a Soviet-backed Indochinese bloc as a direct threat to its security. This shift in national-security perspectives has made possible new forms of cooperation between Indochina and ASEAN.

III. INTERNATIONAL SECURITY: NEIGHBOURING STATES

In the course of the Third Indochina War, differing security perspectives evolved among the six ASEAN nations regarding Soviet-backed Vietnam and its intentions in the region. Thailand and Singapore took a hardline anti-Vietnamese stance and supported Chinese policy. Malaysia and Indonesia were less supportive of the Chinese position and were more sympathetic to Vietnam. Vietnam, for its part, attempted to take advantage of these divisions.

The mistrust nurtured at this time has not entirely dissipated, as demonstrated by contrasting Vietnamese and Thai policies towards Cambodia since 1989. Vietnam boldly reassessed its national-security policies and swiftly disengaged from Cambodia. Thailand, in contrast, has not been able to act. Thai military and civilian leaders are still suspicious of Vietnam and regard it as a threat. This is evident in continued Thai assistance to the Khmer Rouge over four years after Vietnam's withdrawal.

These legacies and present suspicions may restrict the transition of Indochina into full partnership with its South-east Asian neighbours. Foremost among contemporary security concerns are conflicting territorial claims in the South China Sea, border demarcation and control, and the resettlement of refugees and other displaced persons. This chapter also discusses new forms of accommodation and cooperation that are being worked out.

Maritime Disputes

The South China Sea

According to President of the Philippines Fidel Ramos, the potential for conflict in the South China Sea ranks as 'the greatest threat to peace in South-east Asia'.[1] The South China Sea is semi-enclosed, and contains two major archipelagos, the Paracels and the Spratlys.[2] The Spratly archipelago covers an area of over 250,000km^2 and is made up of more than 230 barren islets, reefs, sand bars and atolls, of which about 180 have been named. Roughly 20 of these are technically islands, although few have the fresh water necessary to sustain human life.

The Spratly Islands are of strategic interest for three reasons: they are rich in marine resources; they may contain large undersea deposits of oil and natural gas; and they lie on lines of communication passing through the South China Sea between the Indian and Pacific oceans (on which Japan depends for 70% of its imports). The Chinese Minister of Geology and Mineral Resources estimated that oil reserves around the Spratly Islands could be as high as 30 billion tonnes.[3]

Inter-state relations in South-east Asia acquired new maritime dimensions as a result of the 1982 UN Convention on the Law of the Sea (UNCLOS), which introduced new concepts in ocean law, like EEZs and the doctrine of the archipelagic state. All littoral states in South-east Asia have declared EEZs of 200 nautical miles and thus have laid claim to all the seas within the South-east Asian region; there is no 'free space' in it. Probably 90% of the living resources and almost all of the exploitable non-living resources are found within the 200-mile EEZs. The EEZs of most bordering states overlap, thus giving rise to boundary delimitation issues. The situation is complicated by competing claims to sovereignty to numerous small islands in the South China Sea.

Three nations – the People's Republic of China, the Republic of China (Taiwan) and Vietnam – lay claim to both archipelagos and the entire sea area. Six nations dispute the sovereignty of island territory in this region; China has historically claimed about 80% of the South China Sea. The Spratly Islands are occupied by the military forces of five countries: Vietnam, China, the Philippines, Malaysia and Taiwan. The Paracel Islands are administered and occupied by China.

Since the early 1980s, all ASEAN nations, as well as China, have embarked on naval modernisation programmes, mainly by acquiring fast patrol craft armed with surface-to-surface missiles and other ocean-going vessels such as corvettes and frigates.[4] A variety of factors account for this modernisation, including a reaction to the former Soviet naval presence at Cam Ranh Bay and the desire to enforce sovereignty rights over EEZs. Chinese naval assertiveness prompted Malaysia to occupy two atolls in the southern reaches of the South China Sea in November 1986. A year later a bill was tabled in the Philippines Congress redefining its sea boundaries to include 60 islets in the Spratly archipelago.

By the late 1980s, maritime issues had become a major security concern of the South-east Asian states, including Vietnam.[5] By that time, China had embarked on a naval modernisation programme that brought Chinese hydrographic vessels and warships into the South China Sea on a sustained basis for the first time.[6] Chinese construction crews landed on Fiery Cross Shoal in the Spratly Island archipelago in early 1988 and began building a barracks, wharf, helicopter pad and hangar, and dredging a ship channel.

Almost immediately, Chinese and Vietnamese ships became involved in a number of minor incidents. The Vietnamese Navy was no match for the Chinese Navy, however, and the Chinese expelled Vietnamese vessels from the area. On 14 March 1988, this culminated in a naval engagement close to Johnson Reef in the Spratly Islands in which one Vietnamese boat was sunk and two damaged. It was reported that seven Vietnamese seamen were killed and 74 were missing. At the time of the clash there were an estimated 20 Chinese and 30 Vietnamese vessels of all sizes in the area. In the aftermath of the clash, China extended its control over an additional seven islands. Vietnam responded by reinforcing and fortifying its possessions and by occupying four islets including Barque Canada Reef, also in the Spratlys.

The situation in the Spratly archipelago then remained relatively quiet until 1992, when China once again began to assert itself.[7] On 25 February, China promulgated a law on territorial waters that reiterated its claim to the entire South China Sea and reserved the right to use force in the area to prevent any violation of its sovereignty. That same month, Chinese troops physically took possession of Da Ba Dau (Three Headed Rock), which was unoccupied.[8] On 8 May, China granted a concession for oil exploration to the Crestone Energy Corporation in the disputed area of Vanguard Bank on Vietnam's continental shelf. In early April, Chinese troops planted a territorial marker on Da Lac reef.

Chinese actions provoked regional misgivings. On 22 July 1992, the ASEAN foreign ministers, meeting in Manila, issued a statement calling on unnamed parties to the conflict in the South China Sea to exercise restraint. The ASEAN statement was clearly aimed at China, and it was issued just after Vietnam had acceded to the ASEAN Treaty of Amity and Concord. ASEAN's statement on the Spratlys, and subsequent pressures mounted by individual members, put China on the defensive.[9]

China continued to act in a manner designed to underline its claims. In February 1993, a Chinese oil exploration ship sailed within 20nm of Da Nang. In May another Chinese ship, the *Fendou-4*, entered Vietnam's territorial waters and conducted seismic tests along the central coast that interfered with ongoing Vietnamese exploration.[10] The Chinese ship reportedly ignored demands by Vietnamese escort vessels to leave the area.

In response to Chinese actions, Vietnam reinforced its claims in the Spratly Islands by a variety of means, some of which China saw as provocative. For example, Vietnam has publicised visits by delegations led by senior government officials, members of the National Assembly, scientific survey teams and the Ho Chi Minh

Communist Youth Group. In June 1994, Vietnam's National Assembly formally adopted the 1982 UNCLOS.

According to a Vietnamese political source:

> We have a lot of islands, but up to now they haven't been occupied. We had sovereignty in principle but no people. That left open the opportunity for them to be occupied by others. Now our policy is to have a presence on all of them. Not a military presence – we don't want to shock the Chinese. Economic means are more subtle.[11]

These 'economic means' include construction of physical infrastructure such as fishing ports, piers, solar power generators and TV relay stations and lighthouses, as well as granting tax exemptions to individuals and businesses that exploit and export sea products. By the end of 1992, according to Western intelligence sources, Vietnam had occupied 21 islands, China and the Philippines eight each, Malaysia three and Taiwan one. All territory capable of settlement in the Spratly archipelago was now fully occupied.

The Philippines and Malaysia have both taken steps to assert sovereignty over the islands they claim by garrisoning and fortifying them. Malaysia has built a tourist complex and a short landing strip on one island. In May 1993, the Philippines ordered that an airstrip on Pagasa Island be widened. Over a year later, after protesting to Vietnam about the construction of a lighthouse on Song Tu Tay island, Foreign Minister of the Philippines Roberto Romulo held talks with his Vietnamese counterpart, Nguyen Manh Cam, in Bangkok. They agreed to set up a joint bilateral commission to oversee scientific and environmental studies and the feasibility of joint air and sea rescue operations. Romulo even proposed sporting events between troops. In July 1994, the Philippines also granted a non-exclusive geophysical permit to a US–Philippines joint venture for research in the Reed Bank.

In October 1993, Vietnam and China reached a landmark agreement on the principles for settling of territorial disputes. Both sides agreed not to use force or the threat of force to resolve territorial disputes and to refrain from any action that would worsen the situation. These principles have been endorsed by their respective foreign ministers (July 1994) and deputy foreign ministers (August 1994). During the November 1994 visit to Vietnam by the Chinese President and party head, Jiang Zemin, the two sides agreed 'to form an expert group to deal with issues involving the seas to hold dialogues and consultations'.[12]

Nevertheless, in 1994 China and Vietnam both asserted their sovereignty over the Vanguard Bank area, which heightened regional tensions. On 18 April, the Crestone Energy Corporation announced it was commencing drilling operations in its lease area. Vietnam claimed that this area was not part of the Spratly Islands and was in fact located on Vietnam's continental shelf; it issued an exploration licence to Mobil Oil for the Thanh Long (Blue Dragon) block in an adjacent area. A Vietnamese rig, the *Tam Dao*, began drilling operations shortly thereafter.

In July 1994, on the eve of the 26th ASEAN Ministerial Meeting, China claimed that Vietnamese vessels, including oil-exploration ships, had violated Chinese territory in the Vanguard Bank. Further, Vietnam was charged with repeatedly harassing Chinese survey and fishing vessels in the same area. China announced that two of its ships had turned back a Vietnamese supply vessel, preventing it from reaching the *Tam Dao* rig, while others protected drilling operations by the Crestone Corporation. The following month, Vietnam reported that its Navy had escorted a Chinese research vessel from the Vanguard Bank into international waters.

Prior to 1994, the Spratly Islands question had been primarily a bilateral issue between Vietnam and China, although other claimants were involved. China has been careful not to antagonise other regional states and, indeed, has offered to set aside sovereignty claims and jointly to exploit the resources of disputed territory with them, Vietnam included. ASEAN has chosen a multilateral solution. At its 1994 ministerial meeting, China reiterated its opposition to any efforts to multilateralise the issue. Indonesia's attempt to upgrade the informal workshops on the Spratlys that it had hosted for a number of years was stillborn due to Chinese opposition. In deference to Chinese wishes, the Spratly Islands were not mentioned among the eight points in the Chairman's report on issues discussed at the inaugural meeting of the ASEAN Regional Forum. However, in a 1995 change of policy, China occupied Mischief Reef in the Spratlys, claimed by the Philippines.

Vietnam's impending membership of ASEAN would seem to have implications for resolving conflicting territorial claims in the South China Sea. In terms of *realpolitik*, what is essentially a bilateral dispute between China and Vietnam might turn into a dispute between ASEAN and China. Once Vietnam becomes a full member of ASEAN, China will have to give careful consideration to using force to advance its claims. In such a confrontation Vietnam could well become ASEAN's 'frontline' state, similar to Thailand's position in relation to Cambodia in the late 1970s; (Thailand had been the frontline state when Vietnamese military forces invaded Cambodia and arrived on Thailand's eastern border).

The Gulf of Tonkin

The delineation of the Gulf of Thailand is another maritime concern for Vietnam. Until this problem is resolved, efforts to exploit the hydrocarbon and marine resources of the area will be hampered. This will force Vietnam to divert scarce resources to monitoring and patrolling the areas it claims, and foreign investors are likely to withhold funds until risks are reduced.

China has assertively advanced its sovereignty claims in the Gulf of Tonkin, prompting Vietnam to demonstrate its resolve to protect territorial waters. In August 1992, for example, two Chinese ships erected a drilling platform in a disputed area that the two sides had earlier agreed to leave vacant. China also began impounding Vietnamese ships from Hong Kong, which it claimed were bringing goods to Vietnam to be smuggled into southern China.[13] In August–September 1993, the Chinese ship *Nan Hai Five*, accompanied by two armed helicopters, sailed into waters claimed by Vietnam and engaged in exploration activities before departing. Vietnam made verbal protests and threatened obliquely to use force to expel intruders.

In August 1993, China and Vietnam discussed the situation in the Gulf of Tonkin at high-level talks held in Beijing. After the October 1993 agreement on principles for the settlement of territorial disputes, China and Vietnam announced the formation of two working groups with responsibility for the Gulf of Tonkin and the land border, respectively. Despite continuing discussions incidents still occur. In July 1994, for example, the Vietnamese Navy detained three Chinese fishing boats in the vicinity of Bach Long Vi island, and after being fired on by two other vessels, caught and detained one of them. China protested, claiming that the Chinese vessels were in 'traditional fishing waters'.

The Gulf of Thailand

There have been frequent incidents in the Gulf of Thailand between fishing vessels of the littoral states, in addition to endemic piracy and clashes between government naval forces and smugglers. Thailand is zone-locked in the Gulf of Thailand – it cannot leave its territorial waters and EEZ and enter open seas without passing through maritime territory claimed by another state. Thailand's fishermen have essentially depleted Thai territorial waters. As a consequence, the Thai fishing fleet, one of the world's largest, has been moving farther afield. The search for fish and the need to cut costs have led some Thai vessels to intrude deliberately into the territorial waters of Cambodia and Vietnam.

Disputes between Thai and Vietnamese fishermen are the most serious and often involve the use of automatic weapons and rocket-propelled grenades. Naval vessels from Thailand and Vietnam regularly apprehend boats deemed to be poaching in their territorial waters. Their cargoes are usually confiscated and the crews jailed until they have paid hefty fines. Thailand and Vietnam have been unable to agree on a comprehensive fishing treaty that would regulate the activities of private fishing vessels; until they reach agreement on some regulatory measures, incidents can be expected to continue.

Vietnam has also tried to diffuse tensions between itself and other ASEAN states with which it has overlapping claims in the Gulf of Thailand. In June 1992, for example, Vietnam and Malaysia signed a memorandum of understanding to conduct joint economic operations in areas of the Gulf where their overlapping claims do not involve third parties.[14] Vietnam and Indonesia are in dispute over their continental shelves in the area of the Natuna Islands. Discussions have been under way for a number of years to settle this matter.[15]

Border Demarcation and Control

All three of the Indochinese states face security problems arising from disputes with their neighbours over border demarcation and border control. The total amount of land in dispute is comparatively small; in most cases the question of border delineation is a technical issue, such as locating the exact placement of border markers. However, settlement of border disputes is complicated by issues of sovereignty and national prestige.

The Sino-Vietnamese Border

The Sino-Vietnamese border was the site of a bitter frontier war in February–March 1979. In 1987, both sides reached a *modus vivendi* and permitted cross-border trade. Since then both sides have withdrawn sizeable military forces and have deactivated mine-fields. In 1992, four border posts were officially opened, raising the total number of official crossing points to five; 31 unofficial crossings were reported to be operating.[16] These new border posts linked Yunnan with Lao Cai, Ha Giang and Cao Bang. One consequence has been the phenomenal growth of cross-border smuggling. The volume of traffic has reached such proportions that the flow of illegal goods into Vietnam is seriously undermining local industry. While high-level discussions have been held between senior officials of both governments, particularly since political relations were normalised in late 1991, little tangible progress appears to have been made to resolve a number of outstanding petty territorial disputes. In May 1992, for example, Chinese workers began constructing a railway signal box at Friendship Pass 500m inside territory claimed by Vietnam. This incident delayed the opening of a rail link between the two countries. Shortly after, Chinese workers resumed road building in the same area, 200m inside what Vietnam claimed as its territory. Vietnam later asserted that the Chinese moved the boundary marker at Friendship Pass 400m inside Vietnam.

In early June 1992, Chinese border guards from Guangxi province were accused of evicting Vietnamese farmers from disputed territory and burning their huts. Although Vietnamese officials denied it at the time, it was later revealed that a brief gunfight had broken out between border guards.[17] According to Vietnamese sources, during the first half of 1992 alone, the Chinese military also seized 8,400 hectares of land at 15 disputed locations along their border.[18]

The Laos–Thai Border

Demarcation and control problems along Laos' border with Thailand have been a constant source of inter-state tension since the establishment of the LPDR. In late 1987, differing interpretations of maps based on the 1907 Franco-Siam Treaty resulted in a bitter three-month border war over disputed hills in which both sides suffered several hundred casualties. Thailand closed its border with Laos and imposed an embargo on strategic goods.

In the 1990s, relations between Thailand and Laos have been gradually improving, and in April 1994, the Australian-built trans-Mekong Friendship Bridge was opened. Nonetheless, suspicions were evident until the opening ceremony when the bridge was handed over to Lao–Thai joint ownership. The two could not agree on the precise location of the border: Thailand argued that it should be indicated on the bridge above the deepest channel in the river, while Laos insisted it should be half-way across the bridge. As relations improved, both sides have agreed to open more official border-crossings and to ease restrictions on travel by residents of the border area.

The key diplomatic mechanism for border control and demarcation is the joint border committee formed by Laos and Thailand at national and provincial levels. Laos has been particularly concerned about the cross-border operations of armed insurgents and the possible negative results of opening further bridges across the Mekong. Both sides have expressed concern over drug trafficking, gun-running, smuggling, prostitution and the spread of AIDS. While no agreement at the national level has been reached by these committees on outstanding border demarcation issues, substantial progress has been made at the local level in addressing the concerns of each party.

The Thai–Cambodian Border

During Vietnam's occupation of Cambodia, base camps were established along the Thai-Cambodian border by anti-Vietnamese resistance groups for operations into Cambodia. Western Cambodia has long been viewed by elements of the Thai military and national-security élite as a Thai 'sphere of influence', if not a buffer zone against Vietnamese influence. The Thai military therefore developed close political and logistics relations with the resistance groups. Close economic relations were also established between Thai business interests and the Khmer Rouge, especially in the Pailin area. Thai Army engineers, operating under the UNTAC umbrella, engaged in road repair and construction in the west in order to improve access from Thailand. As late as 1993, an estimated 100,000 Thais still lived in western Cambodia, employed in the timber and gem industries.

After the UNTAC-supervised elections of May 1993, the new Cambodian government, partly comprising former resistance fighters, voiced its suspicions that Thai support for the Khmer Rouge was continuing. These charges were made in particular as a result of the Preah Vihear military debacle, and eventually led to the cancellation of a visit by the Thai Prime Minister. At the end of the year, the question of continuing Thai support for the Khmer Rouge was raised when Thai police intercepted a lorry loaded with weapons heading for the Cambodian border. Thai police also uncovered a major warehouse complex where Chinese-supplied arms were stored.

Refugees

The upheavals in Indochina in the mid- to late-1970s, which culminated in the Third Indochina War, created massive numbers of refugees and economic migrants. During this period, an estimated one million Vietnamese fled their homeland for the West. Another one-quarter million, mainly Hoa people, are currently living in southern China. While the ethnic Vietnamese community in Cambodia, numbering several hundred thousand, has historical ties with the country, an indeterminate number are recent arrivals who left Vietnam for economic reasons. Tens of thousands of Vietnamese who were sent to the former Soviet Union and Central Europe as guest-workers still live there (95,000 in the former German Democratic Republic alone). Over two million Vietnamese live overseas.

Some 300,000 Lao fled the country after the LPRP took power in late 1975 and introduced harsh economic policies. This number includes H'mong tribesmen who had fought against the Pathet Lao and sought refuge in Thailand after 1975. The war in Cambodia also generated several hundred thousand refugees, most of whom fled to Thailand.

The massive outflow of people from Indochina generated a variety of international responses. Only small numbers of Lao and Cambodian refugees have been resettled outside the region, compared to ethnic Vietnamese. The international community, under the auspices of the United Nations High Commissioner for Refugees (UNHCR), drew up a comprehensive programme of action in the late 1980s to govern the resettlement and/or repatriation of Vietnamese who had fled abroad. The UNHCR is also overseeing a resettlement programme of Lao refugees from Thailand – 17,000 have returned home from camps in Thailand, China and other countries since 1980. Over 40,000 Vietnamese have participated in the 'voluntary repatriation programme' from Hong Kong which began in September 1991. Most of the Cambodian displaced persons and refugees in Thailand were repatriated following the 1991 peace agreement.

An estimated 84,000 Vietnamese refugees and displaced persons reside in overseas camps: just over half are in holding centres in South-east Asia, and the remainder in Hong Kong. About 25,000 Lao live in Thai camps, including 22,000 H'mong. An upsurge of fighting in the post-electoral period in Cambodia caused another massive wave of displaced persons.

The question of repatriation has moved up the agenda of Indochina's neighbours. China, Hong Kong, Indonesia, Malaysia and the Philippines have repeatedly pressed Vietnam to accept all refugees for resettlement. Indonesia has taken steps to close the refugee processing centre on Galang Island. Thailand likewise has pressed the Lao government to accept all refugees.

Migration within Indochina has also created tensions. Vietnamese settlers have reportedly moved into southern Laos. While the issue has been kept quiet by both governments, it is of major concern to the Lao that their border be respected. In Cambodia, violence instigated by the Khmer Rouge against the ethnic Vietnamese before the elections in 1993 caused an estimated 20,000 to flee to Vietnam. Eventually Vietnam closed its border leaving 5,000 of them trapped just inside Cambodia; they have remained in UNHCR-supervised camps for over two years. The status of ethnic Vietnamese residents was raised in September 1994 when Cambodia adopted an immigration law denying residence to foreign nationals without proper documentation.

Regional Military Relations

The ending of Soviet military assistance to Indochina exacerbated problems faced by Vietnam (and Laos) as they carried out a major programme of strategic readjustment. Without a modern armaments industry, Vietnam faced acute problems in maintaining its Soviet-supplied modern weapons systems; it took immediate steps to cannibalise or mothball vital equipment. The defence industry was assigned the task of manufacturing whatever spares and replacement equipment it was technologically capable of producing. Research and development was accorded new priority.

In 1991, Vietnam's military establishment consolidated its leadership position within the VCP. The military's increased political influence was apparent in leadership changes announced by the seventh national party congress in June 1991. Two senior generals were appointed to the second- and fifth-ranking positions on the 13-member Politburo. The military representation on the Central Committee increased for the first time since the third party congress in 1960.

Vietnam's military leaders then began to redress the perilous state of the VPA. In 1993, the defence budget rose for the first time since 1987; and it was increased by 50% the following year. Vietnam's military, driven by the need for new sources of vital military equipment and technical assistance (maintenance and refurbishment techniques) also began to look overseas.

Vietnam's developing defence links with its South-east Asian neighbours are part of a larger trend of growing defence contacts with other states. Vietnamese defence officials have visited and discussed various forms of cooperation, training and equipment sales with China, India, Russia, Ukraine, Kazakhstan, North Korea, the Czech Republic, Slovakia, France and Israel. Vietnam and Japan are also considering the exchange of military attachés. Needless to say, contact with Japan was only possible because of a fundamental change in Vietnam's regional military relationships.

Vietnam and Indonesia

In 1991, Vietnamese defence relations with Indonesia took an important step forward. Previously the two nations had maintained contact through delegations at either chief-of-staff or ministerial level. In October 1991, the first contact at staff-college level was initiated with the visit to Hanoi by a delegation from the Indonesian military academy led by its deputy principal. In 1994, this was raised to a higher level when Vietnam received a delegation from Indonesia's National Defence Institute. More significant was the visit of the Assistant Commander-in-Chief of the Indonesian armed forces, Lt-Gen. Teddy Rusdy, later that year. Gen. Rusdy and his group paid a working visit to the VPA's National Defence Industry and Technology General Department to discuss a Vietnamese request for technical assistance, particularly specialist help in the repair and maintenance of military equipment. Vietnam also sought Indonesian involvement in developing its mining sector.[19]

In 1993 the defence ministers of Indonesia and Vietnam paid reciprocal visits. Gen. Murdani visited Hanoi in February, while Doan Khue visited Jakarta in June. After high-level discussions, which included separate meetings with Indonesia's President, Vice-President and Minister for Foreign Affairs, Khue paid a working visit to Surabaya where he expressed interest in the activities of the naval shipbuilding dockyard there, and boarded and inspected a warship. As early as 1990, it was reported that Vietnam wanted to expand its Navy but could not do so because of budget constraints. This was one of the first public indications that Vietnam was interested in purchasing or building corvette-sized naval vessels.

The following year on the eve of the ASEAN Regional Forum, Gen. Edi Sudrajat, Indonesia's new Defence Minister, paid an official visit to Vietnam (10–12 July). Sudrajat met with his counterpart, Gen. Doan Khue, and Gen. Nguyen Thoi Bung, Vice-Minister of Defence. After discussions in Hanoi, Sudrajat flew to Ho Chi Minh City for a visit to the 7th Military Region and a meeting with President Le Duc Anh.

Vietnam and Thailand

In 1990, a year before the Paris peace agreements on Cambodia were signed, Vietnam and Thailand began establishing direct contacts between their military establishments. In March, Gen. Chawalit Yongchaiyut, then Commander-in-Chief of the Royal Thai Army (RTA) and Supreme Commander of the Royal Thai Armed Forces (RTAF), made a one-day visit to Hanoi for discussions with Gens

Doan Khue and Tran Van Quang, both of whom were then deputy ministers for national defence. Gen. Chawalit was also received by Le Duc Anh, then Minister for National Defence, and Do Muoi, Secretary-General of the VCP. In May, Gen. Doan Khue paid a return visit to Bangkok where he held discussions on Cambodia with Gen. Chawalit and other senior RTAF officers. The two generals agreed 'to exchange military attachés, step up exchanges of commanding officers and regularly share military information'.[20]

In June 1991, Gen. Sunthon Khongsomphong, Supreme Commander of the Thai Armed Forces and Chairman of the National Peacekeeping Council, visited Vietnam at the invitation of Gen. Doan Khue. In the course of the visit, Gen. Sunthon held discussions with Foreign Minister Nguyen Co Thach on Cambodia, piracy, overlapping territorial claims, the release of Thai fishermen imprisoned in Vietnam and the exchange of military attachés. Gen. Sunthon reportedly proposed joint naval exercises to combat piracy and to prevent incidents at sea. He also proposed a joint commission to explore and develop resources in areas in the Gulf of Thailand where the two sides had overlapping territorial claims. In August, a delegation from the Thai National Defence College was hosted by the VPA's Military Institute.

In January 1992, Gen. Suchinda Khraprayun, Commander-in-Chief of the Royal Thai Army, paid a three-day visit to Hanoi during which he held separate discussions with Senior Lt-Gen. Dao Dinh Luyen, VPA Chief of the General Staff, Defence Minister Doan Khue, Prime Minister Vo Van Kiet and President Le Duc Anh. Suchinda raised a wide range of issues: assistance in opening a branch of the Thai Military Bank; housing for a Thai military attaché; the establishment of a joint fishing venture to operate in disputed waters; the leasing of Cam Ranh Bay; and the sensitive issue of alleged covert Vietnamese support for refugees living in north-eastern Thailand and alleged Thai support for right-wing anti-communist Vietnamese exiles. He also extended an invitation for Vietnam to send observers to the annual US–Thai *Cobra Gold* military exercises.

Gen. Suchinda further suggested that a hotline be set up between Hanoi and Bangkok to exchange information and to settle disputes as they arose, and that Thailand and Vietnam should exchange military equipment. He proposed that Thailand supply parts from its Chinese T-69 tanks in exchange for parts from F-5 aircraft, which Vietnam had captured in 1975 and still held in storage. Vietnam did not take up the exchange offer and declined the invitation to attend the *Cobra Gold* exercises; it also requested that the Thai Navy refrain from involving itself in the dispute between private fishermen. But Vietnam did express an interest in learning more about the role of the Thai Army in rural development, an item which had not been stressed by the Thais.

The Commander-in-Chief of the Royal Thai Navy led a delegation in June–July 1994 to discuss problems between Thai and Vietnamese fishermen in the Gulf of Thailand. A high-point in developing defence contacts occurred in August 1994 when Defence Minister Gen. Wichit Sukmak visited Hanoi. He met his counterpart, Gen. Doan Khue, and discussed fishery and maritime boundary questions as well as developments in Cambodia.[21] Gen. Wichit was received by President Le Duc Anh. During discussions it was agreed to set up a special commission to assist in resolving fishing disputes in areas of conflicting territorial claims. Khue made a return visit in February–March 1995.

Vietnam and Malaysia

There is no public record of defence contacts between Vietnam and Malaysia prior to the end of the Cambodian conflict. In August 1993, two Royal Malaysian Navy patrol craft paid a three-day port visit to Ho Chi Minh City, the first visit of its kind. This visit may have been made in connection with discussions about the possible sale by Malaysia of patrol boats to Vietnam. It was later reported that the deal fell through because of Vietnam's inability to pay.

In August 1994, a delegation of the Institute of Staff Officers of the Malaysian Armed Forces visited Vietnam for meetings with the VPA's High-Level Military Institute. The two sides discussed staff officer training. The Malaysian delegation was also received by the Minister of National Defence. In late October 1994, Vietnam's Defence Minister Khue headed an eight-member delegation to Malaysia where he held talks with his Malaysian counterpart, Datuk Seri Najib Tun Razak. The itinerary included visits to the staff institute of the Malaysian Armed Forces, Syarikat Malaysia Explosives Technologies, Airod Sdn Bhd in Subang town and the Lumut Naval Base.[22]

The two ministers announced that they would step up the exchange of defence delegations 'to get to know each other better, and to increase mutual confidence'. Minister Najib stated, 'We agree to develop some form of defence cooperation and collaboration, but we didn't go into specifics'. Najib said his remarks were an expression of 'political intention' and the Malaysian minister predicted that there would be 'a great improvement in our military ties' after Vietnam became a full member of ASEAN. 'I prefer them to look at our industry first', he said. Discussions on training were agreed 'in principle' and left for official talks.[23]

Vietnam and the Philippines
Defence contacts between Vietnam and the Philippines have only been made recently. In April 1994, President Fidel Ramos, while on an official visit to Vietnam, offered to make available ten places for Vietnamese cadets at the Philippines Military Academy. He further proposed 'exchanges of visits by senior military officials, study tours for officers and defence instructors and joint ventures in reconditioning of equipment, including aircraft, for re-export'.[24] Vietnamese military officials visited Subic Bay to study its conversion to commercial use in preparation for the possible commercialisation of Cam Ranh Bay. In December 1994, Defence Minister Doan Khue held talks with his counterpart, Philippines Defence Secretary Renato De Villa. This was the first visit by a Vietnamese Defence Minister since the end of the Vietnam War, and according to press reports, the talks centred on the Spratly Islands.

Vietnam and Laos
Laos–Vietnam relations have been transformed in the 1990s. Laos has pressed Vietnam to maintain the spirit and substance of the 'special relationship' proclaimed in the mid-1970s. Hanoi decided not to 'live in the past' and since 1987, has made determined efforts to play down the military and security dimension of the relationship.[25]

Defence contacts between Vietnam and Laos continue to be close despite the withdrawal of VPA units in late 1988. Both countries maintain an active exchange of senior defence officials, including ministerial visits. Relations are particularly close between military departments with responsibility for ideological and cultural matters. In May 1993, the Lao Deputy Minister for Foreign Affairs said:

> The Laos–Vietnam Treaty is still valid. We consider it vital for us too. It is vital in terms of history and neighbourliness and cooperation. We are a landlocked country. But we are not similar to Lesotho. We are not dependent on one country. Improved relations with Thailand and with China are very beneficial for us. Our aim is equilibrium or a balance in our relations. Last year we signed a friendship and cooperation agreement with Thailand.[26]

As a result, Laos–Vietnam relations are now referred to as 'relations of special friendship, solidarity and cooperation'.[27] This choice of words underlines that the period of Vietnamese hegemony in Indochina has ended and Indochina no longer exists as a unified bloc. According to Lao sources, they are considering the implications of letting the 1977 agreement with Vietnam lapse when it expires in 2002. Each state must now determine the pace and scope of its integration within the larger South-east Asian region.

Cambodia

Military relations between Vietnam and Cambodia were terminated with the signing of the October 1991 peace agreements which barred Vietnam – and all other signatories – from providing military assistance. In March 1994, after offensives by the Cambodian Army against the Khmer Rouge's base areas depleted ammunition, a high-level Cambodian delegation led by the two co-ministers of national defence, Tie Banh and Tie Chamrath, was dispatched to Vietnam to request weapons, spare parts and ammunition. According to Norodom Ranariddh:

> we tried to buy ammunition from Vietnam. I sent their old friends to approach the Vietnamese. But President Le Duc Anh wrote an official letter to us saying, 'we are sorry, but we cannot sell you even a single bullet'.[28]

Cambodia has developed modest defence links with states outside the region, like France (training) and Russia (helicopter maintenance). Australia, Belgium, Canada, France, New Zealand, the Netherlands, Norway and the United States have all provided personnel and other assistance to Cambodia's mine-clearance programme. US personnel are also engaged in road building and delivering humanitarian aid. China and Thailand have provided military uniforms. In December 1994, Australia announced a A$5m increase in funding for its defence cooperation programme to Cambodia (for mine-clearance, military communications, naval repair workshop, English-language instruction and financial support for demobilisation).[29] Cambodian requests for arms and ammunition and other so-called 'lethal' aid has met with strong opposition from Thailand, which claims this will exacerbate the situation in Cambodia. ASEAN as a group has adopted a declaration opposing lethal aid. Instead, Thailand has offered to provide training for civic action and, with Indonesia and Malaysia, has offered military training assistance. Cambodia has reportedly taken up the Indonesian offer.[30]

Towards Integration with ASEAN

The end of the Cambodian conflict in 1991 made possible a remarkable transformation of regional security relations. As the Indochinese states carry out their transition to market-oriented economies, they have become more acceptable to the ASEAN states as potential security partners. In particular, Vietnam and ASEAN states have moved quickly to open a dialogue at the military level.

Yet a number of regional security issues have emerged, either as a direct consequence of the decade-long Cambodian conflict, or from having lain dormant for that period. Although many of these issues are capable of resolution through negotiation, the record to date is modest. Suspicions from the past linger. A major question for the future is whether it will be possible to integrate the Indochinese states into a wider regional security framework. From its inception, ASEAN has been reluctant to transform itself into a 'defence community'.[31] ASEAN's defence ties have so far been conducted bilaterally or trilaterally and ASEAN has developed its own unique, low-key style of managing defence relations.

The difficulties of including Vietnam in ASEAN's security community are best illustrated with respect to Thailand. Despite the frequency and level of military exchanges, Thai military officers still regard Vietnam as a security threat. According to Lt-Gen. Poonsak Nakphat, a staff officer for the army commander and Chairman of the National Defence Committee that drafted Thai defence strategies for 1995–99, Vietnam is regarded as a threat because its growing economy would allow it to spend more on defence. If Vietnam improved its defence capability much more, he said, it might try to interfere in Cambodian affairs again. 'The present situation provides no indication that Vietnam would invade Thailand within the five-year period', Gen. Nakphat said, 'but Vietnam is still a potential security threat to Thailand'.[32] He also noted that Thai commercial interests could be adversely affected in the event of a Sino-Vietnamese clash in the South China Sea.

Given Thai reservations (and Singapore's pro-China inclinations), a key question is what form Vietnam's integration into a South-east Asian security community will take. Will Vietnam's participation be limited to a formal level, or will Vietnamese defence officials be warmly received in defence discussions? Will Vietnam be satisfied with ASEAN's slow dialogue process, based on extensive person-to-person contact? Will Vietnam's entry into ASEAN dilute the cohesion of the organisation by provoking an adverse Thai reaction to the development of close defence links with other member states? If, as has been suggested, the present accommodation between ASEAN and Indochina is an insufficient basis for a new regional order, what new forms of security architecture will emerge? Vietnam is a member of the ASEAN Regional Forum, yet the pattern of its relations with major external powers, such as China and the United States, are far from settled.

IV. INTERNATIONAL SECURITY: ASEAN AND THE ASIA-PACIFIC

In the late 1980s, Laos and Vietnam both adopted liberal foreign-investment laws and encouraged direct capital investment from the free-market economies of the Asia-Pacific region. In 1991, party congresses held in both Laos and Vietnam endorsed an 'open door' foreign policy and the necessity of 'making friends with all countries'. For example, Vietnam's *Strategy for Socio-economic Stabilisation and Development Up to the Year 2000*, which was adopted by the seventh party congress, declared that Vietnam would 'diversify and multilateralise economic relations with all countries and economic organisations'.

This marked a major shift in Vietnamese and Lao thinking on the nature of inter-state relations. The transition was not smooth, as conservatives did not want to jettison ideology completely. According to one former party insider, in late 1992 Vietnam still used ideological considerations in setting foreign-policy priorities in confidential internal party documents.[1] Vietnamese officials admit difficulties in adopting this new approach. According to Deputy Foreign Minister Vu Khoan, 'The philosophy of the policy of diversification and multilateralism is that it does not advocate gatherings of force by political systems, but it is defined by national interests and geopolitical and geo-economic origins'.[2]

The October 1991 Cambodian peace agreement and the collapse of the Soviet Union in December represented a major watershed in the development of an 'omni-directional' approach to foreign policy. The Cambodian settlement meant that Vietnam was no longer an international pariah state subject to an aid and trade boycott. After the Cambodian settlement, Vietnam rapidly normalised relations with the countries of South-East and East Asia. Vietnam initially gave priority to the ASEAN states before turning its attention to South Korea, Western Europe, Australasia, Japan and, most recently, Central Europe. In 1994 alone, Vietnam received presidents and prime ministers from Mongolia, the Philippines, Singapore, Sweden, South Korea, Japan, India and Canada.

In 1991–92, Vietnam restored relations with the individual members of ASEAN and with ASEAN as a regional organisation. This was a complete reversal of a policy of confrontation with ASEAN. Vo Van Kiet, then Chairman of the Council of Ministers, led a high-level government delegation to Indonesia, Thailand and Singapore in October–November 1991. The following year he visited Malaysia, the Philippines and Brunei. Kiet's trips marked an effort by Vietnam to return to the regional fold. Since Kiet's visits, Do Muoi, Secretary-General of the Vietnam Communist Party, has paid visits to Singapore, Malaysia and Thailand. He is scheduled to visit Australia and New Zealand in 1995.

The end of the Cambodian conflict brought with it an end to ASEAN's trade and aid embargo against Vietnam. This led to unprecedented levels of commercial interaction. ASEAN investment has increased tenfold from 1991–94 and now makes up 15% of total direct foreign investment. ASEAN states are involved in over 147 projects with a paid-up capital of US$1.4bn by the first half of 1994; 37 agreements have been signed between Vietnam and ASEAN.

Of Vietnam's foreign trade, 60% is with ASEAN states. In 1994, Singapore overtook Japan to become Vietnam's greatest trading partner. Four of the ASEAN countries rank among the top 15 foreign investors in Vietnam. Singapore and Malaysia rank sixth and seventh, respectively, behind Hong Kong, Taiwan, South Korea, Australia and France.[3]

Since 1992, Vietnam has taken part in six ASEAN committees and five ASEAN projects on functional cooperation, including science and technology, environment, health services, population, tourism, culture, civil aviation and maritime transport. In 1992 and 1993, Vietnam and Laos attended the annual ASEAN Ministerial Meetings as observers. In July 1992, both acceded to the ASEAN Treaty of Amity and Cooperation. Two years later, at the ASEAN Ministerial Meeting in Bangkok, ASEAN officially invited Vietnam to become its seventh member and Vietnam's application was formally approved in late 1994. Vietnam will join ASEAN officially on 28 July 1995.

The 1994 Ministerial Meeting made clear ASEAN's desire to see its membership expand to embrace all of South-east Asia. Laos, because it lacks human and financial resources, has decided to go slow on the membership question. Cambodia has been reluctant to go beyond observer status although there are signs that a more positive re-evaluation is under way.[4] In the meantime, both Laos and Vietnam have participated as founding members of the ASEAN Regional Forum.

Laos

In Laos, high priority was assigned to repairing Lao–Thai relations. In March 1991, Gen. Suchinda Kraprayon, Commander of the Royal Thai Army, ended Thailand's estranged relations with Laos by visiting Vientiane. In January of the following year, Kaysone Phomvihane went to Thailand as a guest of the King. An agreement was subsequently reached between the two governments not to interfere in each other's internal affairs; thereafter the threat of domestic insurgency in Laos declined markedly.[5]

Since 1990, Laos has placed priority on reaching agreement on border delineation and control measures with its neighbours. Laos is primarily concerned with preventing disputes over undemarcated territory from flaring up into armed conflict. In 1992, Laos and Myanmar began the difficult task of physically demarcating their common border, a process that was completed in late 1993. Discussions with Thailand, however, have stalled over minor conflicting territorial claims along the border. Laos, recognising the Cambodian government's preoccupation with internal security matters, has not pressed Phnom Penh to open discussions on the delineation of their common border.

Wider East Asia

The collapse of the Soviet Union and the Council for Mutual Economic Assistance (COMECON) system forced Vietnam to reorient its economy to the Asia-Pacific region. Hong Kong, Taiwan and South Korea are now the three largest foreign investors in Vietnam (see Table 1). Japan and Hong Kong are its second and third most important trading partners; two-way trade with each exceeds US$1bn per year.

Vietnam opened diplomatic relations with South Korea in 1992, and they have since exchanged prime ministerial visits. South Korea provides a modest technical assistance and training programme valued at US$50m and, in August 1994, Vietnam and South Korea signed an agreement on cultural and educational exchange. South Korean trade and investment figures are more substantial and South Korea maintains a huge trade surplus with Vietnam. In 1993, South Korea sold goods valued at US$730m while importing only US$90m. South Korean investment, involving 85 projects valued at US$782m, is concentrated in hotel construction, steel mills and manufacturing plants. Vietnam pays close attention to the South Korean model of economic development, which has led to the successful industrialisation of an agriculturally based economy. Vietnam is particularly keen to study South Korea's business conglomerates.

Until recently, the development of Vietnam–Japan relations has been constrained by the commercial debts of Vietnamese state en-

Table 1: **Foreign Investment in Vietnam, as of 11 August 1994.**

Country	Projects	Total Approved Capital (US$)
Hong Kong	204	1.75bn
Taiwan	152	1.71bn
South Korea	85	782m
Australia	45	760m
France	63	739m
Singapore	69	578m
Malaysia	27	559m
Japan	59	517m
Britain	17	402m
Netherlands	13	380m
Switzerland	16	247m
Thailand	52	220m
Russia	49	174m
Indonesia	11	161m
United States	16	159m
Overseas Vietnamese	37	79m

Source: Vietnam State Committee for Cooperation and Investment.

Note: Overseas Vietnamese totals are also included in national totals.

terprises to Japanese businesses and by the continuation of the US embargo. While the embargo was in force, Japanese firms held back from direct investments, as reflected in Japan's relatively low position on Vietnam's foreign investment ladder (59 projects valued at US$517m). It was only in October 1992 that Japan resumed official development assistance after a 14-year suspension.

In contrast, Japan has been Vietnam's largest trading partner and aid donor. Official trade figures for 1993 indicate that Japan imported goods valued at US$1.2bn (mainly crude oil and wood products), and exported to Vietnam goods worth US$667m (machinery and consumer goods). Japanese aid, including loans and grants, was valued at US$500m in 1993. Fifty Japanese firms have opened representative offices in Hanoi, while twice that number operate in Ho Chi Minh City.

Vietnam's Prime Minister visited Tokyo in March 1993, and his Japanese counterpart reciprocated in August 1994. Five agreements on grant aid, valued at US$73m, were signed; the question of Vietnamese commercial debts, incurred by state trading companies, was also discussed. The Japanese Prime Minister promised to raise the ceiling for trade and investment insurance if Vietnam accelerated repayments.[6] He also supported Vietnam's membership of ASEAN, and requested that both countries hold annual discussions at the vice-ministerial level.

It is uncertain how easily ASEAN and Vietnam will adjust to each other. Vietnam will enter ASEAN as its second most populous yet least developed member. Vietnam may not be able to keep up the pace of structural reform and economic growth necessary to benefit from the ASEAN Free Trade Area (AFTA) at an early date. ASEAN's distinctive informal dialogue and consensus, developed over 27 years, may be approaching a period of strain as Vietnam makes its voice heard.

Vietnamese officials have stated that Vietnam looks to both North-east and South-east Asia for its future – which implies that it is looking beyond ASEAN to membership of the Asia-Pacific Economic Cooperation (APEC). However, Vietnam's request to attend the APEC summit in Jakarta as an observer was turned down by the host, Indonesia. APEC itself has put a three-year moratorium on new membership. Vietnam is in an awkward position as APEC members move towards a free-trade area by the year 2020, especially if Vietnam seeks delayed entry into AFTA.

Normalisation of Relations with China

In November 1986, the fourth national congress of the LPRP (anticipating a similar declaration by the VCP a month later), called for improved relations with China (and with Thailand and the United States). The following month, China's Deputy Minister for Foreign Affairs, Liu Shuqing, paid a five-day visit to Vientiane – the first high-level diplomatic contact since 1979. His Lao counterpart, Khamphai Boupha, made a return visit to Beijing a year later, when the two agreed to exchange ambassadors and to open the border for trade. China also gave assurances that it would not support the activities of anti-LPDR refugees.[7] A formal trade agreement was signed in December 1988.

In 1989, as the socialist state system began to disintegrate, Laos redoubled its efforts to diversify its foreign relations. In October 1989, Kaysone Phomvihane, then Secretary-General of the Lao People's Revolutionary Party, restored ties between the LPRP and Chinese Communist Party (CCP) when he made the first of three visits to Beijing.[8] Four agreements covering consular affairs, visa requirements, cultural exchanges and border control were signed.

Sino-Lao economic and political relations developed rapidly. During 1990, for example, Chinese investments in Laos rose to more than half of all foreign investment.[9] Relations reached an all-time high in December when Laos hosted a visit by Chinese Premier Li Peng. Three economic agreements were signed, including a US$9.2m five-year credit arrangement to finance economic and technical co-operation. In March 1991, when the LPRP held its fifth national congress, the CCP was one of only four fraternal parties invited to attend; the others were the Communist Party of the Soviet Union (CPSU), the VCP and the KPRP. Immediately after the congress, Lao Foreign Minister Phoun Sipaseut visited Beijing to brief Chinese leaders.

In late 1990, it was estimated that the Lao refugee population in southern China numbered about 4,000. Repatriation began in March 1991 and was formalised in an agreement signed in August, under which Laos agreed to accept 100–200 refugees per month. According to figures released two years later, the repatriation rate has averaged about 45 per month with 2,917 remaining.

Laos and China have initiated modest defence and party-to-party contacts. In April 1990, the Chief of Staff of the Lao Army accompanied the party secretary of Vientiane on a visit to Beijing. The Lao Defence Minister, Lt-Gen. Choummali Sai-gnason, accompanied Lao Prime Minister Khamtai Siphandon during his October 1991 visit to China. Laos requested Chinese military assistance, but was informed that military equipment could only be provided on a commercial basis. The following year, China sold two Soviet Mi-8 helicopters and some ammunition to the Lao Army. In April 1993, LPRP Politburo-member Thongsing Thammavong went to China to exchange experiences on party-building and personnel training.[10] Chinese Defence Minister, Senior Lt-Gen. Chi Haotian, visited Vientiane in May 1993, and exchanged views with his Lao counterpart, Lt-Gen. Choummali Sai-gnason on 'mutual assistance in building and consolidating the armed forces of each country'.[11] Finally, in July 1994, Laos hosted a brief visit by the Commander of the Chengdu Military Region, Gen. Li Jiulong.

Sino-Lao relations have focused on three main areas: border demarcation, commercial relations; and the return of refugees – with priority given to border issues. During an October 1991 visit to Beijing by the Lao Prime Minister, a treaty was signed establishing a joint border commission. The commission completed its work in late 1992; in February 1993 a comprehensive agreement on border delineation was signed during the visit of China's Foreign Minister.[12] In May, the new Lao Minister for Foreign Affairs, Somsavat Lengsavat, went to Beijing where he exchanged the instruments of ratification on the border agreement with his counterpart, Qian Qichen.

Border issues have since been left essentially to province-level discussions. In June 1994, for example, Luang Nam Tha and Yunnan provinces signed an agreement to cooperate in a wide variety of areas (economy, technology, commerce, tourism, culture and communications). Chinese investment and cross-border trade is expected to increase following an investment protection agreement signed in February 1993.

Laos has been cautious, however, about Chinese proposals to open the Mekong River to navigation and to develop land access routes from southern China to Thailand via Laos. Chinese initiatives are part of a larger scheme to develop a 'growth quadrangle' in the region where Myanmar, China, Laos and Thailand share a common border. In July 1994, following discussions by technical delegations from their ministries of communications and transport, China and Laos signed a draft agreement covering the transport of passengers and goods on the upper Mekong.

The disintegration of the socialist state system in Eastern Europe and the Soviet Union provoked debate in Vietnam about its future relations with China. The party leadership was divided about whether or not to try to establish military relations with China. One group, centred in the VPA, advocated going beyond normalisation and re-establishing formal military and security ties.[13] Another group argued that Vietnam should pursue an omni-directional foreign policy and seek good relations with all countries as the best guarantee of its security.

Vietnam's 'pro-China lobby' was ascendant within the VCP after a secret summit of party leaders in Chengdu, southern China, in September 1990. At the seventh national party congress in mid-

1991, the lobbyists engineered the dismissal of the main opponent of their pro-China line – Foreign Minister and Politburo-member Nguyen Co Thach.[14] After the seventh congress, Gen. Le Duc Anh, second-ranking Politburo-member and then Minister for National Defence, went to China to press for normalisation. Agreement was reached and in November 1991 a summit was held in Beijing, ending 12 years of estrangement. At this meeting, Vietnam pressed China to expand the relationship to include security guarantees just short of a formal military alliance. China rebuffed these approaches, stating that the two could be 'comrades, but not allies'.

Since November 1991, Sino-Vietnamese political relations have steadily improved, even against the backdrop of Chinese assertiveness in the South China Sea. Both sides have exchanged visits by various ministers, foreign ministers and premiers.

China and Vietnam signed 14 major agreements touching on aid, trade, investment, commercial arrangements, scientific and technological exchange, and cultural and educational matters in the two years ending November 1993. A framework for bilateral relations was set out in a joint communiqué issued in November 1991 when relations were normalised, and included guidelines for Vietnam's relations with Taiwan, a touchy subject for Beijing. In October 1993, both sides renounced the use of force to settle disputes and pledged not to take actions that would worsen the current situation.

Sino-Vietnamese economic relations have been restored and have gradually strengthened despite a border dispute and occasional incidents in the Gulf of Tonkin. In late 1992, China granted Vietnam interest-free credit worth 80m yuan (US$14.5m), a token amount, but symbolically important. Since the normalisation of relations, officially recorded cross-border trade had increased from US$40m in 1991, to $179m in 1992, $330m in 1993, and $500m by the third quarter of 1994. Vietnam is Guangxi province's second largest foreign trading partner, illustrating the importance of cross-border trade to the local economies concerned.[15]

While Sino-Vietnamese bilateral relations develop, Vietnam must consider the implications of developments in China over the next decade. Vietnam has a stake in the survival of the CCP and China's success in managing both economic growth and domestic diversity. China's success would vindicate and legitimise to Vietnam's transition to 'Market-Leninism'. Vietnam would be badly shaken if China should falter or experience serious internal instability. To a certain degree, Vietnamese regime security is indirectly dependent on the longevity of the CCP regime.

Following the normalisation of party and state relations, the VPA and the People's Liberation Army (PLA) took measured steps to restore relations. In 1992, Vietnam and China exchanged delegations representing the External Relations Department of the VPA and the Foreign Affairs Bureau of the PLA. The question of Chinese arms sales to Vietnam was reportedly raised during the visit of the PLA delegation in May.[16] According to a Russian diplomatic source, China later provided modest amounts of equipment.

This was followed by reciprocal visits by defence ministers. Gen. Doan Khue visited Beijing in December 1992, where he met his counterpart, as well as CCP Secretary-General Jiang Zemin. Gen. Khue made a point of thanking his hosts for their past 'precious assistance and support'.[17] China's Defence Minister, Sr Lt-Gen. Chi Haotian, reciprocated with a nine-day visit to Vietnam in May 1993. The joint communiqué issued at the end of his visit stated that 'friendly relations between the two countries are historic and should be developed'.[18]

In April 1994, Vietnam's Chief-of-Staff, Sr Lt-Gen. Dao Dinh Luyen, met in Beijing with Gen. Zhang Wannian, his PLA opposite number. According to Vietnamese press accounts, the two 'informed each other about the situation of the economic activities of the two armies, [and] discussed and agreed on the continuation and consolidation of the traditional and friendly relations between the two peoples and the two armies'.[19] Gen. Luyen also met Gen. Liu Huaqing, standing committee member of the CCP Politburo and Vice-Chairman of the Central Military Committee (second in the PLA command hierarchy), and Qiao Shi, Chairman of the Standing Committee of the National People's Congress. Liu was quoted as stating, 'Further development of bilateral and military cooperation is in the interest of both our countries'.[20] Qiao Shi noted that China 'was willing to strengthen cooperation in various fields with Vietnam' and that 'differences existing between the two countries can gradually be narrowed through consultations on an equal footing'.[21]

Sino-Vietnamese relations are by nature asymmetric. Vietnam's population ranks it as a middle-sized Chinese province. The collapse of the Soviet Union and the drawdown of US military power in the region leave Vietnam with few options but to improve relations with its northern neighbour. Vietnam and China have agreed not to let territorial claims and disputes over sovereignty interfere with the conduct of normal relations. Both have pledged not to use force or the threat of force in matters under dispute and not to take any action that would worsen the situation.

Vietnam has sought to maintain its independence by diversifying its foreign relations and by turning to the ASEAN states. However, membership in ASEAN does not guarantee that Vietnam's relations with China will become easier. ASEAN stands to gain little if Vietnam becomes a new front-line state.

Cambodia

Vietnam and ASEAN's problem is eased by China's changed role in Cambodia. Throughout the decade-long Cambodian conflict, China was a consistent supporter of Norodom Sihanouk and the Khmer Rouge. As a result of the UN-brokered peace settlement in Cambodia, China terminated its military assistance to the guerrilla group and formally recognised the coalition government that emerged after the May 1993 elections. While China has distanced itself from the Khmer Rouge, it has continued to support Norodom Sihanouk and international efforts to rehabilitate Cambodia's economy.[22] While China opposes the supply of lethal military aid to the Cambodian government, even to the extent of blocking the private sale of Chinese-manufactured ammunition, it has nonetheless lent diplomatic support to the government in Phnom Penh. In 1994 when the Cambodian government publicly appealed for military assistance, China's provision of uniforms for the Cambodian police was a significant gesture.[23]

Relations with the United States

The United States has maintained formal diplomatic relations with the LPDR since its establishment in 1975, in contrast to Vietnam and Cambodia. American law prohibits the US government from extending any financial assistance to the Lao regime. This law is now under review. A modest humanitarian aid programme is run by the US military as an adjunct to its efforts to recover the remains of servicemen listed as missing in action (MIA) during the Vietnam War. US officials express frustration at the Lao government's reluctance to accept offers for funded visits to the United States or scholarships to US educational institutions. In their view, Laos is wary of dealing with a superpower and finds it easier to deal with middle powers like Australia and Malaysia. Diplomatic relations were raised to the ambassadorial level in August 1992.

The United States first recognised the Cambodian government after the formation of the Supreme National Council. This recognition was then transferred to the Cambodian government after its creation in 1993. The United States opposes, and has threatened to withhold aid from a government of national reconciliation that includes the Khmer Rouge. At the same time, the United States has declined to provide 'lethal aid' to the Cambodian military. This has led to Cambodian complaints that its efforts to deal with the Khmer Rouge are being thwarted.

The United States maintains no diplomatic relations with the Socialist Republic of Vietnam; it applied an aid and trade embargo against North Vietnam from 1964 until reunification in 1975, and then applied existing trade and aid restrictions to the entire country. The US progressively removed restrictions on commercial transactions with Vietnam in 1993. Prior to the lifting of the embargo in February 1994, the United States dropped its opposition to the provision of development assistance to Vietnam by international financial institutions like the IMF, World Bank and Asian Development Bank. Financing from these sources is vital if Vietnam is to rehabilitate and modernise its eroded infrastructure. The two sides have established Liaison Offices in each other's capitals.

The lifting of the embargo led to a marked increase in the activities of US companies wishing to do business in Vietnam, many of whom already had representatives in Hanoi. Despite the initial flurry of activity, US investment remains disappointing to the Vietnamese. The United States ranks fifteenth on the foreign investment ladder with a total of US$159m invested in 16 projects.

Progress towards full diplomatic reorganisation will be determined by the degree of hostility that can be incited by the US domestic MIA lobby, anti-communist Vietnamese-Americans and other relevant human-rights lobbies. These groups wish to link extension of MFN trade status for Vietnamese products, and benefits under the General System of Preferences (GSP), to an improvement in Vietnam's human-rights situation. The United States and Vietnam share a strategic interest: to prevent Beijing from controlling the South China Sea. The US does not officially recognise any historical claims to this area and would oppose the use of force to resolve disputes. Vietnam, in fact, now supports a continuing US military presence in the region. The Commander-in-Chief of US forces in the Pacific hinted at possible strategic interests in Cam Ranh Bay during a visit to Hanoi in late 1994.

Russia

The downward spiral in relations between Vietnam and Russia decelerated in 1992 as Russia and other states of the former Soviet Union addressed the deterioration in economic and commercial relations. Vietnam and Russia set up an inter-governmental Commission for Trade, Economic, Scientific and Technological Cooperation. The Commission has held four meetings with each side represented at the deputy prime-ministerial level. Two-way trade reached a modest US$300m in 1993.

In July 1992, Russian Foreign Minister Andrei Kozyrev announced a complete reversal of Moscow's military disengagement from Vietnam and declared Russia's intention to retain residual forces in Cam Ranh Bay. Cam Ranh houses two important military assets, a functioning signals intelligence station with the capability to monitor Chinese communications around Hainan Island, and a secure harbour where ships crossing the Indian and Pacific oceans can bunker and replenish.

In early March 1994, Russo-Vietnamese military relations were put on new footing with the visit to Moscow by Sr Lt-Gen. Dao Dinh Luyen, VPA Chief-of-Staff. Gen. Luyen held talks with his Russian counterpart, Col-Gen. Mikhail Kolesnikov, about military and technical cooperation. The Russian press reported that this included 'the questions of maintenance, modernisation and production of military equipment' and that 'in the very near future an agreement on bilateral military and technical cooperation will be signed'.[24]

Gen. Dao Dinh Luyen also visited Kiev where he held discussions with the Chief of the General Staff of the Ukrainian Armed Forces and First Deputy Minister of Defence on cooperation in officer training and 'in the technical sphere'.[25] Vietnam is now negotiating barter and off-set agreements with Russia and Ukraine to cover Vietnamese debt repayments, service fees for the use of facilities at Cam Ranh Bay, and Vietnamese purchases of military equipment and access to military training establishments.

Vietnam's political and economic relations with Russia, Ukraine, Belarus, Kazakhstan and Uzbekistan improved with the October 1993 visit of Vietnam's Foreign Minister Nguyen Manh Cam. A high point was reached in June the following year when the Vietnamese Prime Minister made groundbreaking visits to Ukraine, Kazakhstan and Russia. Prime Minister Vo Van Kiet headed a large delegation of his Foreign Minister, 13 deputy ministers or deputy chiefs of state organisations, the VPA Deputy Chief-of-Staff and at least 20 businessmen.

The most significant achievement of Kiet's visit to Moscow was the signing of a new friendship treaty to replace the 'corner-stone' agreement of 1978. According to a senior Russian official, it provides for 'bilateral consultations in the event of a crisis'.[26] The signing ceremony on 16 June 1994 was delayed for several hours because of disagreements over issues like the future of Cam Ranh Bay and Vietnam's desire to renegotiate the terms of the lease granting Russia access to it.[27] Russia's Duma and Vietnam's National Assembly have both ratified the agreement, although the text remains unpublished.

Russian interest in prolonging its stay at Cam Ranh Bay (and Russian suggestions that its facilities be converted for commercial use) raise problems for Vietnam. Vietnamese leaders are searching for a strategic policy towards China. They are concerned about Chinese naval modernisation and provocative actions in the Gulf of Tonkin and Spratly Islands; but they are doubtful of receiving support from Moscow. For example, Vietnamese sources claim that during the March 1988 clash between Chinese and Vietnamese naval forces, the Soviet Union did not fully share the intelligence it gleaned from its signals intelligence facilities.[28]

Vietnam is also anxious to ensure a continued supply of military equipment and spare parts for its armed forces at affordable prices. Russia and Ukraine would be the obvious sources. Moscow has pointed out that continued Russian military cooperation is essential if Vietnam is to maintain and strengthen its armed forces. Vietnam's response was that any future Russian aspirations for a regional role are tied to access to Cam Ranh. The issue is further complicated by Vietnam's indebtedness to Russia and its inability to pay in hard currency.

Vietnam no longer views Russia as a counterbalance to China or, indeed, as a major power in the South-east Asian region. While Russia is no longer a strategic partner, continued access to its military equipment and technology remains important. So too is the economic and commercial relationship. Russia is Vietnam's thirteenth largest foreign investor; the Far East imports foodstuffs from Vietnam, and tens of thousands of Vietnamese continue to work in the former Soviet Union. Also, as a member of ASEAN, Vietnam may help to develop the region's relationship with Russia, given its familiarity with the Russians and the economic ties between the two states.[29]

V. BEYOND INDOCHINA

Communist forces took power in Indochina over 20 years ago. Yet the communist victories of 1975 did not result in a stable 'red brotherhood' of like-minded states. Instead, Indochina became a zone of conflict leading to the Third Indochina War. As a result, South-east Asia was polarised along ideological lines and external powers intervened.

The onset of this war coincided with the failure of the socialist central-planning model and led to 'bottom-up' pressures for reform in Vietnam and Laos. These tendencies were reinforced by Gorbachev's reform programme in the Soviet Union. Domestic and external pressures for change triggered further reforms, which in turn led to a transition from central planning to a market economy.

External relations have also been transformed, particularly after the collapse of communism in Eastern Europe and the Soviet Union. The Indochinese bloc was weakened and then dissolved. Each of the three Indochinese states finds itself in a new international environment where relations among the great powers are fluid and where the boundaries between South-east Asia and North-east Asia are rapidly eroded by the ebb and flow of capital, goods and services. In this new world, Vietnam is poised to become a major actor: it will add 'ballast' to ASEAN in the establishment of a new regional order, but it will also to help to transform ASEAN.[1]

Vietnam formally remains a Marxist-Leninist one-party state whose leaders adhere to the model of mono-organisational socialism. However, in reality the scope of socio-economic change, especially at grass-roots level, effectively means the demise of the mono-organisational model. In the 1990s, the people of Vietnam are freer to go about their daily lives without reference to the Vietnam Communist Party than at any time in the past four decades. Increasingly, the pressures for change are coming from the bottom up, and the Vietnamese state is working out new forms of accommodation with emerging social forces. Through the public security apparatus, the Vietnamese state can repress armed opposition and most organised political dissidence. Yet in many respects, the most powerful pressures for change and reform are coming from within the Party itself where dissidents have mounted powerful intellectual critiques of Marxism-Leninism. The Party has responded by elevating 'the Thoughts of Ho Chi Minh' alongside Marxism-Leninism as its guiding principles, although it is generally accepted that Ho's thoughts enable any party leader to find justification for any policy.

Marxism-Leninism is becoming an icon of the Vietnamese revolution. It is assuming a status akin to *Panca Sila* (the Five Principles) in Indonesia, and it is a symbol of the struggle for national unity. As a result, ideology is fading as a legitimising device. The Party's authority rests on performance, measured by successful economic growth and development. The future of Vietnam depends in part on the ability of the Vietnam Communist Party to adapt to the socio-economic changes. As economic change reshapes Vietnam's political economy, the Party is forming alliances with local officials and entrepreneurs. Vietnam's one-party regime will be transformed into yet another version of 'soft authoritarianism'.

By contrast, the Lao state does not face the same intense pressures for change as Vietnam. This is partly due to Laos' predominantly subsistence economy, the small size of the Lao élite, and limited socio-economic evolution. No major groups have emerged to challenge the LPRP's monopoly on power. In time, Western-educated technocrats will begin to exert a stronger influence on policy, and arbitrary rule will slowly disappear. This is in contrast to the situation in Vietnam, a larger and stronger state that penetrates society to a greater degree than it does Laos, and where the political culture is of a different nature. As a consequence, the LPRP will become more open in its decision-making. It too can expect to remain in power for the foreseeable future.

Cambodia is an exception; the Cambodian state is 'weak' in every sense of the term. The present government is an unstable coalition and the two primary ruling parties, especially Funcinpec, are subject to factionalism. Funcinpec members are increasingly disaffected by their leader, Norodom Ranariddh, and the Party may well be disintegrating. Reformers in the central government have been unable to impose their authority on provincial government, which is largely in the hands of the CPP. The military, while also divided, is a potentially powerful political force. Cambodia's future can only be portrayed as bleak; factionalism may lead to splits within the major parties and the dissolution of the coalition government; the death of King Norodom Sihanouk could spark instability and violence. Cambodia could well revert to authoritarian one-party rule.

External backers of the Cambodian government may try to halt a return to traditional politics, but if political instability reaches a critical stage, they may opt to back a political strongman to bring order out of chaos. The CPP, with its control over the military and its cadre network in the provinces, is likely to become the core of any future Cambodian government. The choice of a CPP leader to become the national leader will depend crucially on support from the armed forces.

The Cambodian government also faces a significant threat from the Khmer Rouge and its guerrilla forces, who are well entrenched in their traditional bases in the north, north-west and south-west. The Khmer Rouge can be expected to persist as a significant menace for at least the next five years. Thus Cambodia's national-security concerns are internal and pressing. But for all the obvious concern about the future of Cambodia, there is relatively little prospect for Cambodian problems infecting the rest of South-east Asia.

As the states of Indochina grow closer to ASEAN, it would be remarkable if ASEAN were not transformed in the process. The decomposition of 'Indochina' as a region is made possible by the revitalisation of Vietnam and, to some extent, of Laos as well. Vietnam can be expected to help force ASEAN states to rethink their policies on everything from regional economic integration to policies towards China, and especially its territorial claims.

The post-Cold War balance of power in South-east Asia has shifted from one dominated by external powers and their alliance network to a more fluid balance in which indigenous actors now play more important roles. Strategic analysts suggest that the new power balance will be determined by the three Asian great powers: China, Japan and India.

The balance of power in South-east Asia is in the process of being reshaped by the trend towards regionalism in reaction to Chinese assertiveness. However, it is unlikely that an expanded ASEAN including Vietnam, Laos, Cambodia and Myanmar, would be sufficient to stabilise a new balance. Regional order in South-east Asia will not be determined by its members alone. It must be combined in a new security architecture that embraces the wider Asia-Pacific region.

Notes

Introduction

[1] Australian Defence White Paper 1994, *Defending Australia* (Canberra: AGPS, 1994).
[2] Michael Leifer, 'Indochina and ASEAN: Seeking a New Balance', *Contemporary South-east Asia*, vol. 15, no. 3, December l993, pp. 269 and 279.
[3] William S. Turley, 'Vietnam's Strategy for Indochina and Security in South-east Asia', in Young Whan Khil and Lawrence E. Grinter (eds), *Security, Strategy, and Policy Responses in the Pacific Rim* (Boulder, CO: Lynne Rienner Publishers, 1989), pp. 166–71; Joseph J. Zasloff, 'Vietnam and Laos: Master and Apprentice', pp. 37–62; Nayan Chanda, 'Vietnam and Cambodia: Domination and Security', pp. 63–76; Macalister Brown, 'The Indochinese Federation Idea: Learning from History', pp. 77–101; and Hollis C. Hebbel, 'The Special Relationship in Indochina', in Zasloff (ed.), *Postwar Indochina: Old Enemies and New Allies* (Washington DC: Center for the Study of Foreign Affairs, Foreign Service Institute, United States Department of State, 1988), pp. 103–14.
[4] Grant Evans and Kelvin Rowley, *Red Brotherhood at War: Indochina Since the Fall of Saigon* (London: Verso, 1984), pp. 1–33 and 165–201.
[5] Leifer, 'Indochina and ASEAN', p. 274.
[6] See Carlyle A. Thayer, 'Indochina', in Ramesh Thakur and Thayer (eds), *Reshaping Regional Relations: Asia-Pacific and the Former Soviet Union* (Boulder, CO: Westview Press, 1993), pp. 201–2 and 220–22; Börje Ljunggren, 'Market Economies Under Communist Regimes: Reform in Vietnam, Laos and Cambodia', pp. 43, 54–58; and Ljunggren, 'Concluding Remarks: Key Issues in the Reform Process', in Ljunggren (ed.), *The Challenge of Reform in Indochina* (Cambridge, MA: Harvard Institute for International Development, 1993), pp. 375–77.
[7] For highlights of forestry products, mining and petroleum and hydropower, see Georgina Carnegie, *Investment Guide to the Lao PDR* (Sydney: Market Intelligence (Asia), 1994).
[8] William R. Thomson, Vice President of Operations, Asian Development Bank, 'Investment Opportunities in the Mekong Region', address to Indochina and Beyond Forum, Sydney, 15 December 1993, p. 8.

Chapter I

[1] It is beyond the scope of this chapter to provide a comprehensive analysis of these complex issues. The following works cover various aspects of them: Carlyle A. Thayer, *Vietnam*, Asia–Australia Briefing Papers, vol. 1, no. 4 (Sydney: The Asia–Australia Institute, University of New South Wales, 1992); Michael C. Williams, *Vietnam at the Crossroads*, Chatham House Papers (London: Pinter for The Royal Institute for International Affairs, 1992); Mya Than and Joseph L. H. Tan (eds), *Vietnam's Dilemmas and Options: The Challenge of Economic Transition in the 1990s* (Singapore: Institute of South-east Asian Studies, 1993); Ljunggren (ed.), *The Challenge of Reform in Indochina*; Gareth Porter, *Vietnam: The Politics of Bureau-*

cratic Socialism (Ithaca, NY: Cornell University Press, 1993); William S. Turley and Mark Selden (eds), *Reinventing Vietnamese Socialism: Doi Moi in Comparative Perspective* (Boulder, CO: Westview Press, 1993); Mio Tadashi (ed.), *Indochina in Transition: Confrontation or Co-prosperity* (Tokyo: Japan Institute of International Affairs, 1989); Melanie Beresford, *Vietnam: Politics, Economics and Society* (London: Pinter, 1988); Martin Stuart-Fox, *Laos: Politics, Economics and Society* (London: Francis Pinter, 1986); Michael Vickery, *Kampuchea: Politics, Economics and Society* (Boston, MA: Allen and Unwin, 1986); and David G. Marr and Christine P. White (eds), *Postwar Vietnam: Dilemmas in Socialist Development* (Ithaca, NY: South-east Asia Program, Cornell University, 1988).

[2] Stefan de Vylder and Adam Fforde, *Vietnam: An Economy in Transition* (Stockholm: Swedish International Development Authority, 1988), pp. 67–71.

[3] Bernard Funck, 'Laos: Decentralization and Economic Control', in Ljunggren (ed.), *The Challenge of Reform*, p. 123.

[4] Ljunggren, 'Market Economies Under Communist Regimes', p. 86.

[5] *Ibid.,* p. 88.

[6] *Ibid.,* p. 52. Factor markets for housing, labour and foreign currencies are underdeveloped.

[7] For overviews of the political reform process, see Carlyle A. Thayer, 'Political Reform in Vietnam: *Doi Moi* and the Emergence of Civil Society', in Robert F. Miller (ed.), *The Developments of Civil Society in Communist Systems* (Sydney: Allen and Unwin, 1992), pp. 110–29; Turley, 'Political Renovation in Vietnam: Renewal and Adaptation', in Ljunggren (ed.), *The Challenge of Reform*, pp. 327–47; and Gareth Porter, 'The Politics of "Renovation" in Vietnam', *Problems of Communism*, vol. 39, no. 3, May–June 1990, pp. 80–91.

[8] Nicholas D. Kristof, 'China Gropes Toward Market-Leninism', *International Herald Tribune*, 7 September 1993, pp. 1 and 4.

[9] 'Political Crisis in Poland', *Nhan Dan*, 25 August 1989.

[10] Radio Vientiane, 7 February 1990.

[11] Martin Stuart-Fox, 'Laos 1991: On the Defensive', and Carlyle A. Thayer, 'The Challenges Facing Vietnamese Communism', in Daljit Singh (ed.), *South-east Asian Affairs 1992* (Singapore: Institute of South-east Asian Studies, 1992), pp, 168–73 and 359–64, respectively.

[12] The author was an official observer at Vietnam's 1992 national elections. This account draws from Thayer, 'Political Reform in Vietnam', *Current Affairs Bulletin* (Sydney), vol. 70, no. 1, June 1993, pp. 22–26; and 'Recent Political Developments: Constitutional Change and the 1992 Elections', in Thayer and David G. Marr (eds), *Vietnam and the Rule of Law* (Canberra: Department of Political and Social Change, Research School of Pacific Studies, The Australian National University, 1993).

[13] These were: Funcinpec, Cambodia People's Party, Buddhist Liberal Democratic Party (BLDP) which won ten seats, and Molinaka (National Liberation Movement of Cambodia), which won one seat.

[14] David Marr, 'The Vietnam Communist Party and Civil Society', Paper to Vietnam Update 1994, The Australian National University,

Canberra, November 1994, p. 14.
[15] Börje Ljunggren, 'Beyond Reform: On the Dynamics between Economic and Poltical Change in Vietnam', Paper to Vietnam Update 1994, The Australian National University, Canberra, November 1994, p. 53.

Chapter II

[1] A point first made by Gen. Vo Nguyen Giap, *Nhiem Vu Quan Su Truoc Mat Chuyen San Tong Phan Cong* ('The Military Mission in Transition to the General Offensive and Uprising') (Uy Ban Khang Chien Hanh Chinh Ha Dong, 1950), and since reiterated. See Gen. Hoang Van Thai, 'Ve Quan He Hop Tac Dac Biet Giua Ba Dan Toc Dong Duong' ('On the Special Relations of Cooperation between the Three Indochinese Peoples'), *Tap Chi Cong San* (January 1982), pp. 17–24.
[2] Grant Evans and Kelvin Rowley, *Red Brotherhood at War*.
[3] Carlyle A. Thayer, 'Vietnamese Perspectives on International Security: Three Revolutionary Currents', in Donald H. McMillen (ed.), *Asian Perspectives on International Security* (London: Macmillan Press, 1984), pp. 57–76.
[4] See Carlyle A. Thayer, 'Indochina', in Desmond Ball and Cathy Downes (eds), *Security and Defence: Pacific and Global Perspectives* (Sydney: Allen and Unwin, 1990), pp. 398–411.
[5] Carlyle A. Thayer, 'The Soviet Union and Indochina', in Roger E. Kanet, Deborah Nutter Miner and Tamara J. Resler (eds), *Soviet Foreign Policy in Transition* (Cambridge: Cambridge University Press, 1992), pp. 236–55.
[6] Carlyle A. Thayer, 'Sino-Vietnamese Relations: The Interplay of Ideology and National Interest', *Asian Survey*, vol. 34, no. 6, June 1994, pp. 513–28.
[7] Gen. Doan Khue, 'Understanding the Resolution of the Third Plenum of the VCP Central Committee: Some Basic Issues Regarding the Party's Military Line in the New Stage', *Tap Chi Quoc Phong Toan Dan*, August 1992, pp. 3–15 and 45.
[8] According to Australian Foreign Minister Senator Gareth Evans, in July 1994 it was estimated that the Khmer Rouge's 'area of control' was 5–10% of Cambodia and its 'area of control and influence' was 10–20%. 'Cambodia: The Current Security Situation', Parliamentary Statement, Canberra, 29 November 1994, p. 4.
[9] Resolution 2 was elaborated on in a series of confidential directives issued by the VCP Central Committee's Military Commission, the Ministry of National Defence and the VPA General Staff's General Political Directorate.
[10] See 'Political Report of the Central Committee (6th Term) at the 7th National Congress', in *Communist Party of Vietnam 7th National Congress Documents* (Hanoi: Vietnam Foreign Languages Publishing House, 1991), pp. 130–33.
[11] Carlyle A. Thayer, The Vietnam People's Army Under Doi Moi, Pacific Strategic Paper no. 7 (Singapore: Institute of South-east Asian Studies, 1994).
[12] For background see Martin Stuart-Fox, 'Socialist Construction and National Security in Laos', *Bulletin of Concerned Asian Scholars*, vol. 13, no. 1, 1981, pp. 61–71; Stuart-Fox, 'National Defence and Internal Security in Laos', in Stuart-Fox (ed.), *Contemporary Laos: Studies*

in the Politics and Society of the Lao People's Democratic Republic (St Lucia: University of Queensland Press, 1982), pp. 220–44; Stuart-Fox, *Laos*, pp. 136–43; and Stuart-Fox, *Vietnam in Laos: Hanoi's Model for Kampuchea*, Essays on Strategy and Diplomacy no. 8 (Claremont, CA: The Keck Center for International Strategic Studies, Claremont McKenna College, May 1987).

[13] See Carlyle A. Thayer, 'Laos and Vietnam: The Anatomy of a "Special Relationship"', in Martin Stuart-Fox (ed.), *Contemporary Laos: Studies in the Politics and Society of the Lao People's Democratic Republic* (New York: St Martin's Press, 1982), pp. 245–73.

[14] 'Political Report of the Lao People's Revolutionary Party Central Committee', General-Secretary Kaysone Phomvihan, Fifth Party Congress, Vientiane, 27 March 1991, in *Pasason*, 28 March 1991, pp. 3–9.

[15] *Ibid.*

[16] The three governments are: Front Uni National Pour un Cambodge Indépendent, Neutre, Pacifique et Cooperatif (Funcinpec); KPNLF; and the Party of Democratic Kampuchea (PDK, Khmer Rouge), led by Prince Norodom Ranariddh, Son Sann and Khieu Samphan respectively. People's Republic of Kampuchea, State of Cambodia, Supreme National Council, Provisional National Government of Cambodia and finally Kingdom of Cambodia.

[17] See *Cambodia's New Deal: A Report by William Shawcross*, Contemporary Issues Paper no. 1 (Washington DC: Carnegie Endowment for International Peace, 1994), p. 72.

[18] National Voice of Cambodia, radio broadcast, 27 July 1993. Senator Gareth Evans, 'Cambodia: The Current Security Situation', Parliamentary Statement, Canberra, 29 November 1994, p. 6. According to Western intelligence sources, at the time of the merger, ANKI and the Khmer People's National Liberation Front (KPNLF) totalled 10,000 each while the CPAF stood at 90,000 (excluding a police field force of 40,000–45,000) for a total of 110,000. Each of these groups deliberately overestimated their troop strength when reporting to UNTAC. ANKI and the KPNLF provided 'official' figures of 17,400 and 27,500 respectively. The CPAF reported a strength of 131,000 (excluding police), for a grand total of 175,900.

[19] There are an estimated 450,000 weapons in circulation in Cambodia. Estimates for the number of mines range from 4–10m.

[20] Prince Norodom Ranariddh, National Voice of Cambodia, radio broadcast, 8 August 1993.

[21] On 31 March 1994, Melissa Hines, an American employee of the Swiss organisation, Food for the Hungry International, and her Cambodian colleagues were taken hostage and held for 41 days before being released on 11 May. Their ransom was paid in kind – a village well was dug and humanitarian supplies provided. On 12 April 1994 two British citizens and one Australian were kidnapped at Sre Ambel in Kampot province. Although ransom demands were made it appears the hostages were murdered almost immediately. On 26 July three more Westerners – a Frenchman, an Australian and an Englishman – were taken hostage in Kampot

province. They were later executed. This particular case was widely publicised by the foreign press.
[22] John C. Brown, 'RCAP Outlines Reform Plan', *Phnom Penh Post*, vol. 3, no. 17, 26 August–8 September 1994, pp. 1 and 5.
[23] Consideration is being given to allocating provincial governorships on the basis of 11 for Funcinpec and 10 for the CPP.
[24] Cambodian specialists disagree about the extent of factionalism within the CPP. Some argue that the CPP is split into conservative and reformist factions led by Chea Sim and Hun Sen respectively. See David Chandler, *Cambodia*, Asia–Australia Briefing Papers, vol. 2, no. 5 (Sydney: The Asia–Australian Institute, University of New South Wales, 1993), pp. 7–8; and Michael Vickery, *Cambodia: A Political Survey*, Regime Change and Regime Maintenance in Asia and the Pacific Discussion Paper Series no. 14, Department of Political and Social Change, Research School of Pacific Studies, The Australian National University (Canberra), 1994, pp. 35–38.
[25] Shawcross, *Cambodia's New Deal*, p. 72.
[26] Private information, interview, Vientiane, 3 May 1993.
[27] Murray Hiebert, 'Soldiers of Fortune', *Far Eastern Economic Review*, 13 June 1991, pp. 26–27.
[28] Murray Hiebert, 'Taking the Flak', *Far Eastern Economic Review*, 10 November 1988, p. 42.

Chapter III
[1] *Reuters*, Manila, 5 December 1993.
[2] Aprilani Sugiarto, 'The South China Sea: Its Ecological Features and Potentials for Developing Cooperation', *The Indonesian Quarterly*, vol. 18, no. 2, Summer 1990, p. 116. It also contains the Pratas islands and the Macclesfield Bank, which are not discussed here.
[3] *Xinhua*, 5 September 1994; and Graham Hutchings, *Daily Telegraph*, 7 September 1994.
[4] Tim Huxley, *Insecurity in the ASEAN Region*, Whitehall Paper Series (London: Royal United Services Institute for Defence Studies, 1993), pp. 50–71; and Carlyle A. Thayer, *Trends in Force Modernization in South-east Asia*, Working Paper No. 91 (Canberra: Peace Research Centre, Research School of Pacific Studies, Australian National University, September 1990).
[5] Various Vietnamese forces have continually occupied islands in the Spratly group since the 1950s following the departure of the French. Claims by private Filipino citizens to islands in the South China Sea prompted Taiwan to occupy Taiping Island in the mid-1950s. The Philippine government then pre-empted private exploration by declaring sovereignty over the islands immediately adjacent to Palawan province. Minor clashes occurred between Filipino fishermen and Malaysian naval vessels in 1988.
[6] For documentation of Chinese assertiveness in the South China Sea, see John W. Garver, 'China's Push Though the South China Sea: The Interaction of Bureaucratic and National Interests', *China Quarterly*, December 1992, no. 132, pp. 999–1028. See also, Michael Leifer, 'Chinese Economic Reform and Security Policy: The South China Sea Connection', *Survival*, vol. 37, no. 2, Summer 1995, pp. 44–59.

[7] This section is based in part on Carlyle A. Thayer, 'Vietnam: Coping with China', in Dajlit Singh (ed.), *South-east Asian Affairs 1994* (Singapore: Institute of South-east Asian Studies, 1994), pp. 351–67.
[8] Andrew Sherry, *Agence France-Presse*, Hanoi, 16 July 1992. According to Vietnamese sources this was the eighth islet to be occupied by China since 1988.
[9] For example, the matter of Chinese intentions in the Spratlys was raised by President of the Philippine Fidel Ramos during a state visit in April 1993, in discussions when the Chinese Minister of Defence, Gen. Chi Haotian, visited Malaysia (May 1993), and again when Vice-President Joseph Estrada visited Beijing in June 1994. Estrata suggested demilitarisation and joint activities in the Reed Bank area. The third round of informal talks by claimant nations was held in Manila, 31 May–2 June 1993; the fourth round on 23 August; and the fifth round, in October 1994. In July 1994, at the 27th ASEAN Ministerial Meeting, China, Vietnam and the Philippines all had bilateral discussions on the Spratlys. ASEAN and Chinese officials discussed this matter at a meeting in Hangzhou in April 1995.
[10] *Voice of Vietnam*, radio broadcast, 11 May 1993.
[11] Quoted by Andrew Sherry, *Agence France-Presse*, Hanoi, 27 October 1993.
[12] Point 4 of the Sino-Vietnamese joint communiqué issued in Hanoi, 22 November 1994.
[13] The figures on the number of ships vary from 14–20; see *Reuters*, Hanoi, 31 March 1993.
[14] Ismail Kassim, 'Malaysia, Vietnam Agree on Framework for Joint Oil Search', *Straits Times* (Singapore), 6 June 1992.
[15] This section is based in part on Carlyle A. Thayer, 'The Security Dimension of Maritime Issues', in Gehan Wijeyewardene (ed.), *South-east Asian Borders in Regional Context* (Singapore: Institute of South-east Asian Studies, forthcoming 1995), chap. 12; and Michael Leifer, 'Maritime Regime and Regional Security', in Tai Ming Cheung (ed.), *Changing Patterns of East Asian Security* (Hong Kong: Centre for Asian Pacific Studies, Lingnan College, December 1991).
[16] Thayer, 'Vietnam: Coping with China', in Singh, p. 361.
[17] 'Guards Exchange Fire at Border', *Asian Defence Journal*, August 1992, p. 95.
[18] The Sino-Vietnamese border is delimited by 314 markers of which about 70% are still in place. The Vietnamese claim that the People's Liberation Army has moved forward at 142 border markers and occupied 80km^2 of land.
[19] Indonesia's Department of Defence and Security and a private company are reportedly investigating joint investment in Vietnam's coal sector.
[20] Thai Army Television, 28 May 1990.
[21] *Agence France-Presse*, Hanoi, and Kulachada Chaipipat, Hanoi, *The Nation* [Bangkok], 25 August 1994.
[22] Zarina Tahir, 'KL, Hanoi to Step up Defence Cooperation', *Business Times* [Singapore], 2 November 1994; and Ismail Kassim, 'Malaysia and Vietnam to Boost Defence Co-operation', *Straits Times*, 2 November 1994.
[23] Voice of Malaysia external service, Kuala Lumpur, 1 November 1994.

[24] Agence France-Presse, *Straits Times*, 12 April 1994.
[25] Private information, interview with Vietnamese Foreign Ministry personnel, Hanoi, November 1991.
[26] Interview, Ministry of Foreign Affairs, Vientiane, 4 May 1993. Lesotho is a landlocked country entirely surrounded by South Africa.
[27] This was the expression used in the joint communiqué issued after Lao President Nouhak Phomsavan's visit to Vietnam in August 1994.
[28] Quoted in Nate Thayer and Nayan Chanda, 'Things Fall Apart', *Far Eastern Economic Review*, 19 May 1994, p. 17. Ranariddh initially denied that a request for 'ammunition supplies' had been made. Prior to approaching Vietnam, Cambodia had made a similar request to Laos and China. Attempts to purchase Chinese-manufactured ammunition through a Singaporean company was blocked by China.
[29] 'Additional Assistance for Cambodia', Department of Defence and Department of Foreign Affairs and Trade, Joint News Release, M153, Canberra, 29 November 1994.
[30] Based in part on Thayer, *The Vietnam People's Army Under Doi Moi*, pp. 63–73; and Thayer, *Vietnam's Developing Ties with the Region: The Case for Defence Cooperation*, ADSC Working Paper No. 24 (Canberra: Australian Defence Studies Centre, Australian Defence Force Academy, June 1994), pp. 6–18.
[31] See the discussion in Amitav Acharya, 'The Association of Southeast Asian Nations: "Security Community" or "Defence Community"?', *Pacific Affairs*, vol. 64, no. 2, Summer 1991, pp. 159–78.
[32] 'Military Thinks Vietnam May Still Pose Big Security Threat', *The Nation*, 22 September 1994, p. A5.

Chapter IV

[1] See Bui Tin, *Breaking the Ice: The Memoires of a North Vietnamese Colonel* (London: C. Hurst and Co., 1995), p. 191.
[2] Interview with *World Affairs Weekly*, cited in Vu Son Thuy, 'Vietnam and Asean: The Road Ahead', *The Nation*, 19 August 1994.
[3] Data on foreign investment provided by the State Committee for Cooperation and Investment as of 11 August 1994.
[4] According to an Australian diplomat who attended the ASEAN Post-Ministerial Meeting and inaugural meeting of the ASEAN Regional Forum in Bankgok in July 1994, Cambodia's new positive attitude towards ASEAN is a by-product of King Sihanouk's ill-health, as Sihanouk has discouraged Cambodian membership in ASEAN. ASEAN holds over 250 official meetings each year. Laos does not have the manpower to participate fully in them. Private communication, Canberra, September 1994.
[5] According to a Western ambassador interviewed in Vientiane in May 1993, 'There has been a major decline in resistance activity over the last several years. There was a report in March of this year about an attack on Route 7 near Luang Prabang, in which ten persons were killed. This would be the first major report of such activity in over a year. The Thais have made the decision to cut back and limit their assistance to the resistance. There is one Thai Colonel involved in corruption and gun running. But compared to two to three years ago, when support was higher, the present activity is much lower', private communication.

[6] Some of Vietnam's commercial debts date back to the 1980s. Vietnam's failure to settle them has resulted in a long-festering financial problem. In November 1992, a major hurdle was overcome when Vietnam cleared a debt of US$200m with a bridging loan from Japan. Japan then extended new soft-loan credits worth US$400m. In 1993 Vietnam made a debt repayment of US$37m. The total amount of outstanding debt has not been disclosed.

[7] There was an upsurge in anti-LPDR activities in late 1987 by groups based in Thailand. Chinese assurances on this matter must have played a role in the Laos–Vietnam decision to reduce the presence of VPA forces in Laos during 1988.

[8] He also travelled to Japan and France in his first official foray outside the socialist bloc (visits to Thailand excepted). He made return visits to China in September 1991 and May 1992; he died later that year.

[9] *Straits Times*, in *Agence France-Presse*, 26 November 1990.

[10] China is presently training Lao party officials in Yunnan.

[11] Vitthayou Hengsat Radio Network, 15 May 1993.

[12] An initial protocol with map was signed in August 1992.

[13] See the discussion in Thayer, 'Sino-Vietnamese Relations', pp. 513–28, and Thayer, 'Vietnam: Coping with China', in Singh, pp. 351–67.

[14] Carlyle A. Thayer, 'Comrade Plus Brother: The New Sino-Vietnamese Relations', *The Pacific Review*, vol. 5, no. 4, September 1992, pp. 402–6.

[15] Brantly Womack, 'Sino-Vietnamese Border Trade', *Asian Survey*, vol. 34, no. 6, June 1994, pp. 503–5.

[16] Jacques Bekaert interviewed on the BBC, *Dateline East Asia*, 31 July 1992, in Singapore, *Foreign Broadcast Monitor* 176/92, 1 August 1992, pp. 8–9.

[17] *Xinhua*, 8 December 1992.

[18] *Xinhua*, 14 May 1993.

[19] Hanoi Radio, 14 April 1994.

[20] *Reuters*, Beijing, 10 April 1994.

[21] *Xinhua*, Beijing, 11 April 1994.

[22] China's change of policy has been criticised by Khmer Rouge leaders in confidential documents and private remarks. King Norodom Sihanouk, who is stricken with prostate cancer, has been continually treated in Beijing since his illness was first diagnosed.

[23] The section is based on Carlyle A. Thayer, 'Vietnam's Strategic Readjustment', in Stuart Harris and Gary Klintworth (eds), *China as a Great Power in the Asia Pacific* (Melbourne: Longman Cheshire Pty Ltd, forthcoming); Thayer, 'Sino-Vietnamese Relations', pp. 513–28, Thayer, 'Vietnam: Coping with China', in Singh, pp. 351–67; and Thayer, 'China's Domestic Crisis and Vietnamese Responses, April–July 1989', in Gary Klintworth (ed.), *China's Crisis: The International Implications,* Canberra Papers on Strategy and Defence No. 57 (Canberra: Strategic and Defence Studies Centre, Australian National University, 1989), pp. 83–97. For background, see Charles McGregor, *The Sino-Vietnamese Relationship and the Soviet Union*, Adelphi Paper 232 (London: Brassey's for the IISS, 1988).

[24] Mayak Radio, Moscow, 2 and 10 March 1994. Gen. Luyen also met with the First Deputy Minister for Defence, Andrei Kokoshin.

[25] *UNIAR* news agency, Kiev, 9 March 1994.

[26] *ITAR–TASS*, World Service, Moscow, 16 June 1994.
[27] The other issues in dispute included the role of Vietsovpetro in oil and natural gas prospecting and production; problems relating to Vietnamese living in Russia; the role of the UN; and Vietnam's debts.
[28] Author's interview with Lt-Gen. Vu Xuan Vinh, Hanoi, 12 May 1993.
[29] Hoang Anh Tuan, 'What are the Consequences of a New ASEAN Member?', *Straits Times*, 29 August 1994. This section is based on Carlyle A. Thayer, 'Russian Policy Toward Vietnam', in Peter Shearman (ed.), *Russian Foreign Policy* (Boulder, CO: Westview Press, 1995), pp. 201–23; Thayer, 'Indochina', in Ramesh Thakur and Carlyle A. Thayer (eds), *Reshaping Regional Relations: Asia-Pacific and the Former Soviet Union* (Boulder, CO, San Francisco, CA and Oxford: Westview Press, 1993), pp. 201–22; and Ramesh Thakur and Carlyle A. Thayer, *Soviet Relations with India and Vietnam, 1945–1992* (Delhi: Oxford University Press, 1993).
This chapter is based in part on Thayer, *Vietnam*, pp. 55–61. For background, see Amitav Acharya, *A New Regional Order in South-East Asia: ASEAN in the Post-Cold War Era*, Adelphi Paper 279 (London: Brassey's for the IISS, 1993), pp. 41–52.

Chapter V

[1] 'It's a possibility they'll become a full member of ASEAN. If they do ASEAN gets a lot of weight added to it, doesn't it?'. Remark by Australian Prime Minister Paul Keating, quoted by Laura Tingle, 'PM Tells Asia to Get Tough on Trade', *The Australian*, 11 July 1994, p. 2.